The Science of Musical Sound

The Science of Musical Sound

Revised Edition

John R. Pierce

W. H. Freeman and Company
New York

Library of Congress Cataloging-in-Publication Data

Pierce, John Robinson, 1910–
  The science of musical sound—rev. ed.
  Bibliography: p.
  Includes index.
  1. Music—Acoustics and physics. 2. Music—
  Psychology. 3. Sound. 1. Title.
  ML3807.P5 1983 781'.22 82-21427
  ISBN 07167-6005-3

Printed in the United States of America

2 3 4 5 6 7 8 9 0   VB   9 9 8 7 6 5 4 3

To Max Mathews, whose Music V and whose kind and patient counsel started many things in many places.

Contents

Preface

The first edition of this book, published in 1983, was written in a different world. Then, the computer generation of musical sound was a lusty but tiny infant. Commercial synthesizers were chiefly analog devices, costly and tricky. Digitally generated sound, now familiar through inexpensive digital keyboards, was limited by cost and capability.

Today analog sound is obsolete. Audio technology *is* digital technology. Digital keyboards are more common than pianos and are more suited to our limitations of space and money. Composers produce movie and television scores by digital means, without the intervention of instrumentalists.

Musical sound is musical sound, whether it is generated by acoustical instruments or by digital hardware, and musical sound is what this book is about. *The Science of Musical Sound* describes the physical and mathematical aspects of sound waves that underlie our experience of music as well as the psychoacoustics of musical sound — the relation of the physical aspects of sounds to their perceptual features.

The historical and scientific basis of this book isn't all that different from the first edition, for science and sound don't change that rapidly. Technology does; new technology and new work have added to our knowledge, and I have sometimes amplified and sometimes corrected what I said in the first edition. There are other changes in the book. All chapters have been updated but the most heavily reworked are Chapter 1 on the role of computers in music, Chapter 12 dealing with sound reproduction, and Chapter 13 on synthesized sound and equipment for producing it. I have added an appendix on MIDI, that surprising standard through which any commercial keyboard can operate any commercial digital synthesizer, and which in general allows interconnection of digital sound gear of various manufacturers. Andrew Schloss, who urged the republication of this book, suggested this addition and drafted the appendix. It was revised with his aid and that of David Jaffe. Further, David Zicarelli has supplied an appendix on MAX, a popular programming language widely used in the control of MIDI-compatible synthesizers.

I have drastically revised, pruned, and added to a bibliographical appendix. I have cited books and pertinent recordings on compact discs of both digital sound examples and of a very few examples of computer music that make important points.

Sound examples are essential to our understanding of sound. Today, a fair amount of computer music has appeared on compact discs, through Wergo and other publishers. And, there are a few recordings of sound examples which are referred to in the bibliography. Beyond this, many commercial synthesizers are available, and many of these can be used in producing the sine waves and combinations of sine waves described in various chapters. Thus, diligent readers can find sound examples on various discs or, better yet, synthesize them on their own.

The preface to the first edition told how I used the fifth Marconi International Fellowship award in preparing the first edition and of my gratitude to Gioia Marconi Braga, Marconi's daughter, for the opportunity the award honoring her father gave me. It told of my indebtedness to Jean-Claude Risset, to Elizabeth Cohen, and to many at Stanford's CCRMA (pronounced *karma*, the Center for Computer Research in Music and Acoustics) for their work in preparing sound examples. It told of people who, in my division at Bell Laboratories, had inspired my interest in sound — of E. E. David, Jr., and Max Mathews, whose early work, culminating in his Music V program, launched computer music to the world in 1957. Today both Max Mathews and I are professors in the music department of Stanford University, invited to join CCRMA by its director, John Chowning. His invention of fm synthesis made possible the first reasonably priced digital keyboard (Yamaha's DX7).

In the preface to the first edition I mentioned the happy month that I spent at Pierre Boulez's IRCAM (Institute for Research and Coordination of Acoustics and Music) in Paris in 1979. I acknowledged the help and inspiration of many: including the late Gerald Strang, Manfred Schroeder, and Earl Schubert, Jr. And of Gerard Piel, Linda Chaput, and others at Scientific American Library who had made the book possible.

Several colleagues at CCRMA and elsewhere have been helpful in preparing this revised edition. I have already mentioned the contributions of Andrew Schloss, David Jaffe, and David Zicarelli. I have consulted frequently with Earl Schubert. Jay Kadis, CCRMA's audio engineer, was very helpful in connection with Chapter 12, and so was James A. Moorer of Sonic Solutions. Above all, John Chowning, director, and Patte Wood, administrative director, have been essential to this book — as to everything else that happens at CCRMA.

I also wish to thank Linda Chaput, president of W. H. Freeman, for making this new edition possible, and Christine Hastings and others at Freeman for their patient and diligent help in the revision of the manuscript and in seeing it through to publication.

John R. Pierce

1 *Sound, Music, and Computers*

*T*his book is about musical sound. Perhaps we should say *acoustics*, a word that entered the English language in the seventeenth century, the century in which Galileo wrote about musical sound.

Today papers in the *Journal of the Acoustical Society of America* are divided into many categories, some of which have little relation to music, such as "Underwater Sound," or "Ultrasonics, Quantum Acoustics, and Physical Effects of Sound." "Music and Musical Instruments" is a category, but much that is essential to an understanding of musical phenomena appears under "Psychological Acoustics" and "Physiological Acoustics." In today's world, knowledge has become so divided and extended that it sometimes seems hard to make sense of it.

In the Greek and Roman worlds, music, including whatever was known about musical acoustics, held a high place in science and philosophy. In the liberal arts of the Middle Ages, music was a part of the higher *quadrivium*, along with arithmetic, geometry, and astronomy. The place of music in the liberal arts was above that of grammar, rhetoric, and logic, which constituted the lower *trivium* that dealt with words rather than numbers.

As time passed, music's status became more complicated. The romantics tended to associate music with the grandiose sentiments rather than the grand scientific insights of their age. Still, through the nineteenth century, scientists studied music and musical sound with insight as well as aesthetic appreciation. In 1862 Hermann von Helmholtz, physiologist, physicist, great scientist on a grand scale, published *On the Sensations of Tone as a Physiological Basis for the Theory of Music*. Not a word in the

1

title about acoustics, though we would say that the book is about musical acoustics and psychoacoustics and, in addition, about music in general.

In this century some musicians have looked to science and technology for new directions in music without concentrating on the word *acoustics*. These have included Edgard Varèse and Hermann Scherchen. This has not been the chief current of musical thought, nor has music been a part of the mainstream of science. The greatest influence of science on music has come through the development of means for recording and reproducing the sounds of music played on conventional musical instruments. The phonograph, with its later electronic advances, and radio revolutionized the role of music in our lives as radically as photography, motion pictures, and television have changed our world of visual experience. Today the computer and digital technology in general are working fantastic changes in the recording and transmission of sound, and in the generation of musical sounds.

This book is indeed about acoustics — both the physical acoustics pertinent to the understanding of conventional musical instruments and the sounds they produce, and the psychoacoustics that helps us to understand the perception of musical sounds. But it is about acoustics in relation to music and musical ideas.

The changes that computers and their descendants will continue to work in music will come partly through fresh insights of musicians who work with new sounds. But computers have opened up new ways of analyzing and experimenting with sounds, and new ways of investigation of the response of human beings to sound, including musical sounds. Today we know far more about sounds and their perception than we did in the pre-electronic era. And we will know more in the future.

We can be sure from past experience that new sounds and new understanding of sound will affect the course of music profoundly. Better sounds have always produced different music.

Musical instruments have improved greatly in range and quality in the past few centuries, and certainly up to the beginning of this century. In part, this improvement resulted from (1) the development of better, more easily playable instruments, especially brass instruments with valves and woodwinds with better key mechanisms; (2) the increasing skill of instrumentalists; and (3) an expansion of the range of musical sound, as composers and performers developed and exploited new effects and new idioms.

Whether or not we wish to call such change progress, it brought an expansion in the variety of orchestral sound. Think for a moment of the sounds of Bach, Mozart, Wagner, Debussy, and Stravinsky. As successive generations of composers expanded into new territories of concept, orga-

Figure 1-1 Krummhorns, a Renaissance group of double-reed instruments.

nization and sound itself, the music that they produced was different in style and sound. They did not confront the past on its own ground.

In our century, electronic sounds in general have had a profound effect on some nonelectronic music. When Edgard Varèse wrote Déserts in 1954 for taped sound and orchestra, he was proud that he had provided such continuity of sound quality that it is hard to detect transitions from tape to orchestra. Some of Krzysztof Penderecki's music of the 1960s for conventional orchestra deliberately imitates "electronic" sound quality, as does some orchestral writing of Yannis Xenakis a little earlier. In such works, written at a time before synthesized and computer-produced

Figure 1-2 A modern B-flat bass clarinet.

sounds escaped from their early "electronic" timbre, we hear the orchestra refining sounds whose somewhat harsh qualities many listeners found objectionable.

The influence of electronic music on some composers has been more subtle. György Ligeti, who had worked with Karlheinz Stockhausen at the West German radio electronic studios in Cologne from 1958 to 1960, abandoned the limited and difficult electronic means of tone production available at that time. Nevertheless his music for voices and conventional instruments shows that he is acutely aware of the subtle qualities of electronic sounds and of the musical value of the sophisticated control of sound quality.

Figure 1-3   Edgard Varèse (left).

Computers and Music

When, in 1957, Max Mathews first used a digital computer to produce complex musical sounds, it seemed that this could have a liberating effect on composers. In principle, a computer can produce *any* sound. Its potential is limited only by the composer's imagination. But of this Milton Babbitt said, "It's like a grand piano in the hands of a group of savages. You know that wonderful sounds can come from it, but will they?" The computer will produce novel and fine music, but only through human skill and effort.

Today the ubiquitous personal computer can be adapted to generate musical sounds. Inexpensive digital keyboards are more widespread than pianos, and a seemingly endless variety of commercial digital hardware is available for producing, modifying, and analyzing musical sounds. Composers of many different styles of music are making themselves heard in homes, in concerts, and through TV and film scores.

Does the future of musical sound lie in digital synthesis? Will the computer have any influence on other aspects of music, for example form or organization? Some have thought it might.

In 1957 Lejarin A. Hiller, Jr., and his collaborator L. M. Isaacson took their inspiration from Johann Joseph Fux (1660–1741), who codified rules to describe the stylistic practices of earlier contrapuntalists, most notably Palestrina. Hiller and Isaacson programmed a computer to make random choices constrained by some of Fux's rules for first-species counterpoint,

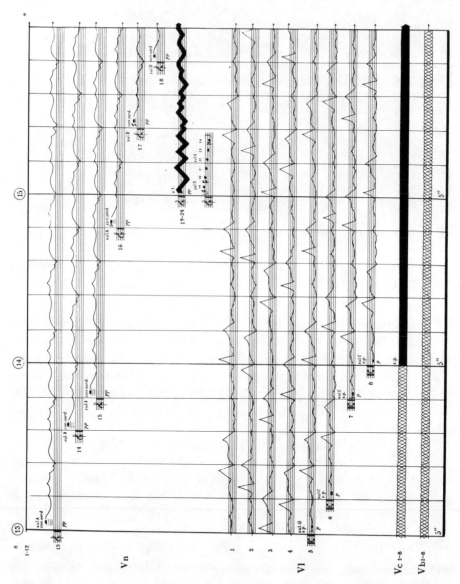

Figure 1-4 A page from the score of Krzysztof Penderecki's *Polymorphia*.

Figure 1-5 Milton Babbitt.

which sets notes of three countermelodies against each note of the original melody. A little of this music, called the *Iliac Suite*, sounds pleasant, but it wanders, and so does the listener's attention. Traditionally, the rules of counterpoint are not intended to tell you what to do, but what *not* to do. The chance element used in generating the *Iliac Suite* provided some surprise but no overall sense of direction.

A lack of musical purpose may be inherent in machine composition. Rules in music are not canned algorithms that we can use in making a

Figure 1-6 Lejarin A. Hiller.

computer solve cut-and-dried problems over and over again. What, indeed, *is* the place of computers in music?

Hiller and others have pursued the idea of the computer as a composer, or at least, as a tool in manipulating musical material. I will say little about the computer as a composer's aid, for that is a subject unrelated to understanding and generating musical sounds.

We may note, however, that some help that the composer needs can be and is supplied by a computer. One of the most useful tasks that a computer can do is to produce musical scores of high quality in a day when even clear hand copies of music have become excessively expensive. Leland Smith's pioneer SCORE program is now available for IBM (or IBM-compatible) personal computers. It produces scores of excellent, publishable quality. It is easy to make changes and to extract instrumental parts from a full orchestral score. The input to SCORE is the computer keyboard. Many later programs have been developed for producing musical scores on both Macintosh and IBM computers, including Finale, Professional Composer, and NoteWriter. Some allow playing on a pianolike keyboard as an input.

The role of the computer in composition goes beyond the production of a neat final score. Computers are used to store and manipulate musical materials, including lists of notes or their equivalent.

I think that the chief challenge of the computer lies in another direction, that of new sounds and their use. In the past, many composers have responded to the challenge of creating new tone colors. Harry Partch invented both a new scale and an entire orchestra of new instruments to play his music, but Partch's instruments were difficult to build and are not commercially available. In their search for new timbres, some composers have evoked strange sounds from conventional instruments. Perhaps the strangest was the sound of a violin burning on a New York stage, an event staged by Lamont Young and Charlotte Moorman.

Two early-twentieth-century analog instruments, the Theremin and the Ondes Martenot, were recognized as unique musical resources by a number of composers who wrote idiomatically for them. We have noted that electronic analog (as opposed to digital) synthesizers appeared in the mid-twentieth century. They played an appreciable role in music through the 1960s. Robert Moog's analog synthesizers had a distinct musical impact, for example, through Walter Carlos' *Switched On Bach.*

Electronics was an essential part of *musique concrète* of Pierre Schaeffer and others at the Studio d'Essai of the French radio system. Analog synthesis was pursued in the West German radio studios at Cologne by Karlheinz Stockhausen and others.

Analog electronics tended to be expensive and not to stay in adjust-

Figure 1-7 Herbert A. Deutch, with the earliest prototype of the Moog synthesizer, which he helped invent.

ment. Therefore, analog synthesis did not have a musical impact comparable to that which digital synthesis has had.

New Technology and Computer Music

The computer offers a wide range of sounds, along with the means for controlling them very accurately. The challenge is how to master a constantly changing medium of unlimited acoustic potential, and how to find aesthetic reasons for realizing these new capabilities.

In the early days of computer music, composers encountered a number of problems. There were no instruments available. Composers had to create their own instruments as computer software. And they had to play them; no performers were available either. The composer had to supply all input through a typewriterlike keyboard. All these factors proved awkward in performances. Once a composition was completed, who wanted to sit in an auditorium and listen to music coming from loudspeakers? An audience could not even be sure when to clap unless the composition gave a clue, or unless the house lights came up. Was there an alternative to the concert? Only a few commercial recordings were made, and none had a wide distribution.

Some of the troubles with concerts were overcome in various ways. One solution was to couple recorded sounds with projected images, as in Andy Moorer's *Lions Are Growing*, a setting of a poem by Richard Brauti-

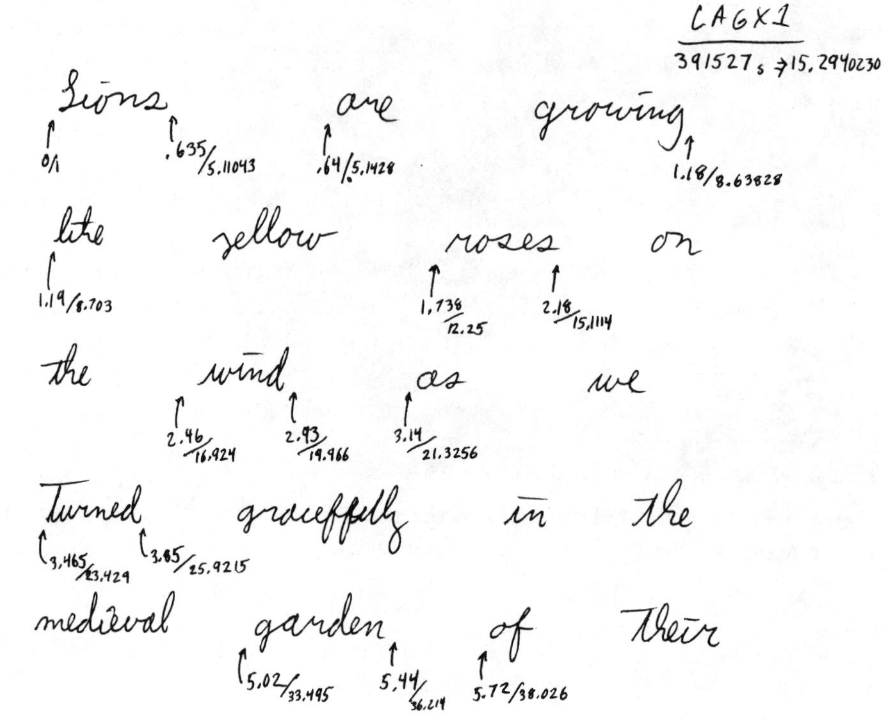

LAGX1

391527$_s$ ⇒15,2940230

Lions   are   growing
↑   ↑   ↑
0/1   .635/5.11043   .64/5.1428   1.18/8.63828

lite   yellow   roses   on
↑   ↑   ↑
1.19/8.703   1.738/12.25   2.18/15.1114

the   wind   as   we
            ↑   ↑   ↑
         2.46/16.924 2.93/19.966 3.14/21.3256

Turned   gracefully   in   the
↑   ↑
3,465/63.429 3.85/25.9215

medieval   garden   of   their
              ↑   ↑   ↑
         5.02/33.495   5.44/36.214   5.72/38.026

roaring   blossoms .   Ooh,   I
↑   ↑   ↑   ↑
6.07/40.292 6.53/43.269   7.25/47.93   8.01/52.85   8.1/54.73

want   to   turn .   Ooh,   I
↑   ↑
8.53/56.216 8.70/57.446   9.23/60.682 10.2/61.026 10.56/69.356 10.59/69.55

on   turning .   Ooh,   I   have
                 ↑   ↑   ↑   ↑
              11.57/75.894 12.7/83.2 12.99/85.046 13/85.15

Turned .   Thank   you
↑   ↑   ↑   ↑
13.49/88.322 14.07/92.077 14.6/95.507   15.0/98.097

Figure 1-8 A working computer score for Andy Moorer's *Lions are Growing*.

gan. In this effective piece composed in 1978, a computer-processed voice speaks, sings one line of music and chords, and roars, accompanied by appropriate slides. This made computer music into a component of a multimedia event, like later digitally synthesized film and TV scores. Today, computer film and TV scores are no more and no less recorded and replayed than are scores played by musicians on conventional instruments.

Another concert alternative has been to couple recorded synthesized sound with a singer or instrumentalist. This can be very successful, but it has not been the only resource. In this day of digital keyboards (Figure 1-10), parts or all of a score of digital sound can be evoked using a conventional keyboard. Through the world-standard MIDI interface, a standard commercial keyboard can control commercial digital synthesis hardware. Or a keyboard player can supplement or interact with or control and modify the process of digital sound synthesis.

Max Mathews's Radio Baton, a sort of computerized musical drum, offers another solution. In all modes of operation the player strikes or

Figure 1-9 Max Mathews and his Radio Baton. Among other things, the Radio Baton enables the performer to beat out the rhythm, loudness, and sound balance of a piece whose notation is stored in computer memory.

strokes a surface with one or more drumsticks. The velocity of striking can control loudness; the position can control timbre in one direction and pitch in another. The successive pitches and durations can also be stored in the computer. Rate of striking or position of stroking can control tempo, and position of striking or stroking can change sound quality and instrumental balance. In this way the Radio Baton brings the performer into the realm of computer-generated sound, but the skills required are perhaps nearer to those of a conductor than of a traditional percussionist. A number of similar devices have appeared.

Today, digital synthesis of sound is used by composers all over the world, in universities, in conservatories, in computer-music institutions, and in commercial music. The MIT press publishes a quarterly *Computer Music Journal*. The worldwide Computer Music Association holds an annual International Computer Music Conference, as well as other meetings, and issues a publication called ARRAY. There are commercial publications, including *Keyboard* and *Electronic Musician*. When I look back over more than thirty years to the time when Max Mathews generated the first computer music piece in 1957 I am struck by obstacles overcome and progress made.

Figure 1-10 The Yamaha SY99, a late model digital keyboard. The earlier Yamaha DX7, which reached the market in 1983, was the first completely digital keyboard synthesizer, and the first such synthesizer that sold at an affordable price (around $2,000). Early "digital" keyboard devices sold for about ten times this price. The success of the DX7 was due partly to Japanese persistence and ingenuity, partly to the use of special integrated circuit chips, and partly to the use of fm (frequency modulation) synthesis, an invention of John Chowning. Now discontinued, the DX7 was a landmark that ushered in a new era. (*Photo courtesy of Yamaha Corp. of America*)

Many talented young composers use computers and digital synthesis hardware as experimental tools to study the intricacies of musical sounds. Some will produce digitally synthesized music. Others will compose music for conventional instruments. All will be influenced, all will learn many new and useful things. And so, I hope, will the reader of this book.

Although this book discusses computers and the digital analysis and synthesis of musical sounds, it is really about the well-known aspects of all aspects of musical sounds, about pitch, scales, consonance, harmony, and timbre, and about some less-known aspects of perception. We can't have a useful understanding of musical sounds without considering these aspects. We will start with periodicity, pitch, and waves.

2 Periodicity, Pitch, and Waves

*A*lthough wind instruments have been known for nearly five thousand years, and harps for almost as long, sounds of a definite pitch are not necessary to music. The earliest musical instruments that archeologists have found in Egypt are clappers. Perhaps song accompanied their rhythmic beat, but the music may have been largely rhythmical. A knowledgeable friend of mine tells me that, in very primitive music, the chief interval used is the fifth (seven semitones), though sometimes an indefinite musical third (four semitones) is also used.

Rhythm by itself can make music. In our time, Carlos Chávez composed a toccata for percussion alone. A siren is heard in Edgard Varèse's *Ionisation*, but that fine work achieves its effect chiefly by rhythm and *timbre* (sound quality).

Pitch and Periodicity

What *is* pitch? Psychologists insist that pitch is a name for our subjective experience of periodic waveforms, rather than a physical property of the sound wave that reaches our ears. Loosely, we can use the word *pitch* to denote the shrillness of a sound. In this sense, the hissing sound *s* has a higher pitch than the shushing sound *sh*. In this chapter, pitch is considered to be a definite quality related to musical tones, such as those produced by the violin, the clarinet, the tuba, the piano, or the human voice. We hear such sounds as having definite pitches that correspond to particular notes of the musical scale. The present *musical* standard for

14

"concert pitch" is that the A above middle C sounds at 440 vibrations each second. More on pitch can be found on pages 36–37.

Sounds that have a definite, unambiguous pitch are called *periodic*, because something happens over and over again at a constant rate. Galileo found by accident that he could produce a sound having definite pitch by scraping a brass plate with a sharp iron chisel. The tiny parallel and equidistant ridges left on the brass were a permanent witness to the vibrations of the screeching chisel that had engraved them. An old book on musical acoustics relates that Galileo also produced a pitched sound by rubbing a knife rapidly around the edge of a milled coin. You can try this by scratching the edge of a quarter with your fingernail. The sound produced as your nail encounters the ridges around the edge of the coin does have some pitch. The faster you scratch the coin, the higher the pitch.

The siren provides a clear illustration of the periodicity related to musical pitch. The very siren that Varèse used in *Ionisation* stood in his New York studio. When I turned the crank faster, the pitch of the sound that the siren produced rose. Why was this? In order to understand, we must examine the mechanism of the siren, which was invested by Charles

Figure 2-1 De la Tour's siren.

Cagniard de la Tour in 1819. Reduced to its essential parts, a siren consists of a rotating disk pierced by a number of equally spaced holes near its circumference, and a nozzle that directs a jet of air through the holes as they pass, as shown in the diagram in Figure 2-2. Such a siren emits a puff of air each time a hole on the disk passes the nozzle.

The graph in Figure 2-3 represents the succession of puffs of air from the siren. As time passes (moving from left to right), there are intervals when no air passes through a hole (the low, flat parts of the curve) and shorter intervals when puffs of air pass through a hole, shown by asterisks where the curve rises and falls. The succession of puffs of air that the siren produces is periodic: That is, successive puffs are produced at equal intervals, T seconds apart. The time T between successive puffs is called the *period*. The number of puffs in a standard unit of time (usually one second) is the *frequency*.

The periodic pulses of air that the siren emits set up a periodic disturbance called a *sound wave* that travels through the air. When it reaches our ears, we hear this periodic wave as a sound with a definite musical pitch. When the siren produces 440 pulses per second, we hear the A above middle C (in concert pitch). If there are 220 pulses per second, the sound is an octave lower. If there are 880 pulses per second, the sound is an octave higher. Figure 2-4 shows the frequency of various notes of the

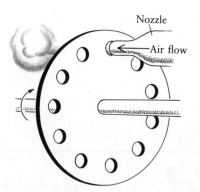

Figure 2-2 A simple siren. Compressed air from a nozzle passes through a circle of equally spaced holes in a rotating disk. As puffs of air pass through the holes, they excite a periodic vibration in the air. The number of puffs per second is the number of revolutions per second times the number of holes in the circle. This disk has 11 holes. If it revolves 40 times a second, it will produce 440 pulses a second, which corresponds to the pitch A above middle C.

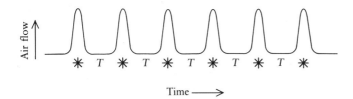

Time ⟶

Figure 2-3 Periodic puffs of air produced by a siren. The time, *T*, between one puff and the next (indicated by asterisks) is called the period.

piano keyboard, together with the ranges of pitch, or *compasses*, of several other musical instruments.

Only sources of sound that are periodic have a clear, definite, unarguable pitch. In this chapter, and in several that follow, only periodic sounds and their pitches are considered, for such sounds form the basis of scales and traditional harmony.

Frequency and Pitch

We might think that the relation between frequency and musical pitch was discovered by means of the siren. The number of pulses of air that the siren produces in a second is the number of holes around the disk times the number of times that the disk revolves each second. We can drive the disk at a high speed by turning a crank attached to a train of gears. If we measure the number of times per second that we turn the crank, we can calculate the number of times per second that the disk rotates, and hence the frequency of the pulses at each pitch that the siren produces.

In fact, the relation between frequency and pitch was discovered much more slowly and much less directly, long before the siren was invented. The pitch of musical sounds that we now know to be periodic was a central aspect of music long before frequency had been related clearly to pitch. Many early cultures used pitch in an orderly and effective way. Like other peoples, the Greeks must have noticed from the earliest times that plucked strings vibrate. Various Greek philosophers associated high notes with swift motion and low notes with slow motion, though they gave no exact relation of motion to pitch. The new discovery that the Greeks made and passed on to posterity is the wonderful numerical relation between the lengths of strings and musical intervals. This discovery is commonly attributed to Pythagoras (c. 500 B.C.)

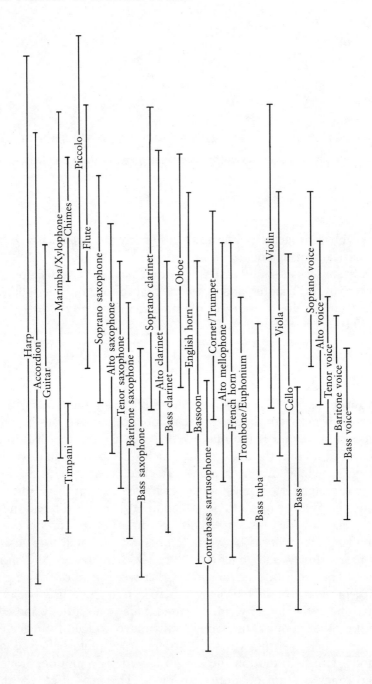

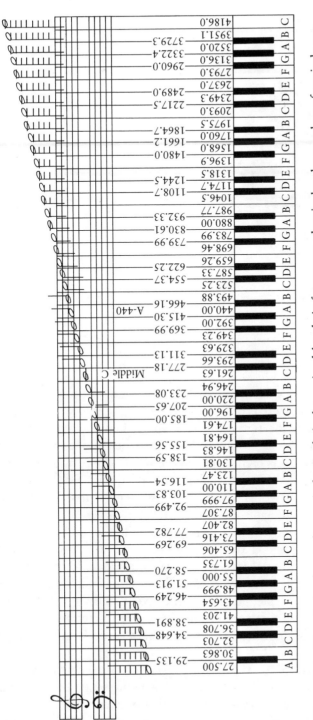

Figure 2-4 The pitch of periodic musical sounds is determined by their frequency, that is, by the number of periods per second. This diagram relates the notes of the musical scale, the positions of the keys of the piano, and the ranges of various musical instruments to the corresponding frequencies. Some periodic sounds have frequencies above or below the range of the piano keyboard. For such sounds, changes in frequency don't correspond to clear musical intervals, though the sensation of pitch does go up and down with frequency. Such sounds don't have a clear or useful *musical* pitch.

Figure 2-5 Greek vase painting of a citharist. The cithara was sacred to Apollo.

Imagine a stretched wire of length L, as shown at the top of the first part of Figure 2-6. If the wire is plucked, it will emit a sound with a definite pitch, say, middle C. If the tension of the wire is kept constant, but the length is shortened by placing a solid wedge somewhere along the wire, the pitch of the plucked wire is higher. The figure shows the relation between the lengths of the plucked wire and the pitches produced. For example, if the wire is shortened to two-thirds of its original length, it will sound the note that is a *fifth* (seven semitones) above the original note.

The Greeks had a rather mystical regard for number and proportion. They were gratified to find a relation between the ratios of the whole numbers and familiar musical intervals. They sought similar harmonious relations for the proportion of buildings. This was characteristic of their thought. Plato, for example, identified the five regular polyhedra (see Figure 2-7) with the four elements and the universe (tetrahedron, fire; cube, earth; octahedron, air; icosahedron, water; dodecahedron, the universe).

To more modern minds the relation between the lengths of strings and musical intervals is empirical, and people sought some physical explanation for it. From the Renaissance on, scientists wanted to get behind the regularities of complex phenomena and find their explanation in simple terms. In *Dialogues Concerning Two New Sciences*, published in 1636,

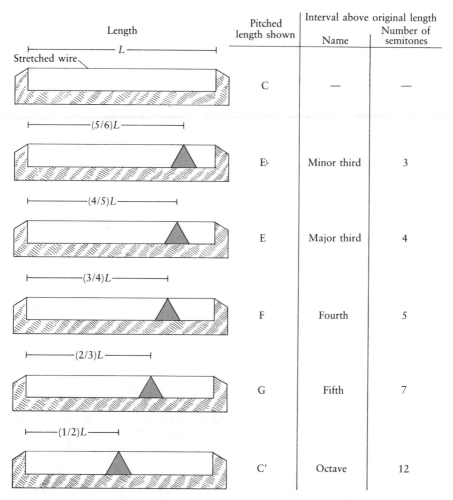

Length	Pitched length shown	Interval above original length	
		Name	Number of semitones
Stretched wire — L	C	—	—
(5/6)L	E♭	Minor third	3
(4/5)L	E	Major third	4
(3/4)L	F	Fourth	5
(2/3)L	G	Fifth	7
(1/2)L	C′	Octave	12

Figure 2-6 A stretched wire vibrates with a particular frequency and gives a tone of a particular pitch. Here we assume this pitch to be middle C. If we keep the tension the same, but use a wedge to reduce the vibrating length to 5/6 of the original length, the frequency increases by 6/5, and the pitch goes up a minor third, to E♭ (E-flat). Other fractional reductions in the length of the string result in other musical intervals: 4/5, major third, E; 3/4, fourth, F; 2/3, fifth, G; 1/2, octave C.

Galileo clearly explained the relation between pitch and the frequency of vibration of a string, but he wrote only of the relative numbers of vibrations per second corresponding to various musical intervals.

In *Harmonic Universelle*, also published in 1636, the French cleric, philosopher, and mathematician Marin Mersenne related pitch to the

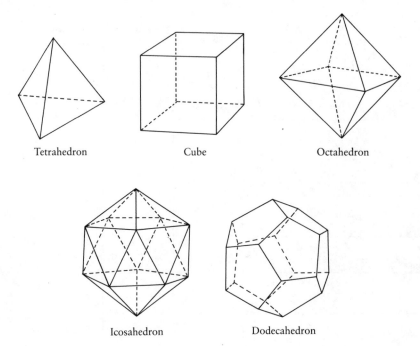

Tetrahedron   Cube   Octahedron

Icosahedron   Dodecahedron

Figure 2-7 The five regular polyhedra.

actual number of vibrations per second. Like Galileo, with whose work he was familiar, Mersenne knew how the frequency of vibration varies with the length of a stretched string (frequency is proportional to the reciprocal of the length*), with tension (it is proportional to the square root of the tension), and with mass per unit length (it is proportional to the reciprocal of the square root of mass per unit length). Putting all this together, we find

$$\text{frequency} = k \times \frac{\sqrt{\text{tension}}}{\text{length} \times \sqrt{\text{mass per unit length}}}$$

But by what factor k must we multiply this product of quantities in order to get the actual number of vibrations per second? Mersenne found the correct factor by counting the number of vibrations per second of long strings, including a hemp cord 90 feet long and $1/12$ inch in diameter, and a brass wire 138 feet long and $1/48$ inch in diameter.

---

*For any number n, the reciprocal is $1/n$.

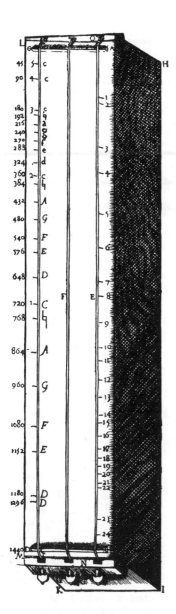

Figure 2-8 The monochord, used to relative lengths to pitch, as depicted in Mersenne's *Harmoni Universelle*.

Not everyone immediately appreciated or accepted these ideas. According to Samuel Pepys's diary entry for August 8, 1666, Robert Hooke told him that "he is able to tell how many strokes a fly makes with her wings (those that hum in their flying) by the note that it answers to in musique during their flying." Pepys characterized this as "a little too much refined." Nevertheless, we now know that the periodic nature of musical sounds arises from the nature of waves, in air, water, and strings.

Figure 2-9　Portraits of Galileo, Mersenne, and Kepler.

Sound Waves

We have all seen the circles of ripples that move outward when a raindrop falls into a quiet pool, or when we drop a pebble into smooth water. In a similar way, a disturbance of the air moves out from the disturbing source. However, sound waves do not travel merely on a surface, but through the air in all directions. The air in a sound wave does not move bodily, as water flows in a stream, but only locally. One part of the air imparts motion to the air ahead, as might happen if, in a long line of closely spaced people, a person at the end gave a push to the one ahead, and that one, in turn, pushed the next. We can imagine a disturbance traveling to the head of the line without anyone taking a step forward.

We experience sound in air, but we can't see sound waves. Furthermore, they pulsate so rapidly that we can't feel their individual pulsations, except perhaps in the lowest notes of a pipe organ. We can visibly represent the motion

Figure 2-10 Ripples on water.

of the air in a sound wave with a contrivance made of a series of little weights connected by springs (see Figure 2-11). If we give a sharp blow to one end of this device, the weights down the line move forward and back in turn, and we see a wave travel along the series of weights and springs. The figure shows successive positions of the masses as a single wave, or pulse, travels to the right. This contrivance produces waves that, unlike sound waves, travel in only one dimension, a straight line. However, the springs and weights accurately represent the two properties of air that allow the propagation of sound waves — elasticity and mass.

We are familiar with the elasticity of air from experience with automobile tires and bicycle pumps. When we compress air, it gets smaller and fits into a smaller volume. Conversely, air under pressure expands if it can, as we see when we blow into a toy balloon. That air has mass is evident in every breeze and strong enough to make visible objects move. Sailboats and the leaves of trees move in the wind because the moving air imparts to them some of its momentum, a property of moving things that have mass.

A single sudden disturbance, such as the explosion of a firecracker, pushes the air next to the disturbing object. Because air has mass and elasticity, it resists and is compressed. The compressed air then expands again, pushing in all directions against the air around it. The surrounding air in turn becomes compressed, forming a shell of compressed air a little

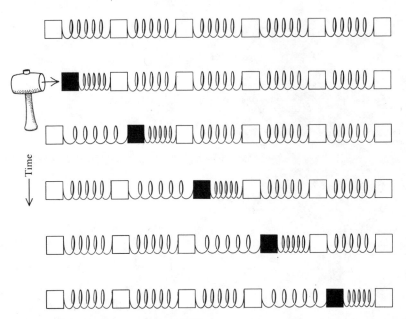

Figure 2-11 A wave can travel along a sequence of weights connected by springs. Waves travel through air in a similar manner, for air has both mass and springiness, or elasticity.

distance from the original disturbance. The expansion of the air in this shell creates yet another shell, farther out, and so on (see Figure 2-12). Thus, we can think of a single sound wave as an expanding shell of compression. Successive layers of air are compressed and decompressed as the wave moves outward from the source of disturbance, but each individual air molecule moves only a little distance out and back. A steadily vibrating object, such as a plucked string, starts an expanding shell of compression with each vibration. Thus, we can visualize the emitted tone as a series of expanding shells (see Figure 2-13).

In Figure 2-11 and in actual sound waves, the movement of the medium, whether springs and weights or air molecules, is back and forth along the direction in which the wave itself travels. Such waves are *longitudinal*. In contrast, waves in water are *transverse*: That is, most of the motion of the water is up and down, *sideways* to the direction of the visible wave. Plucked strings also exhibit transverse waves.

The tension in a stretched string is a force that tends to straighten out bends in the string. Because of the tension, a stretched string, when pulled aside at one point, pulls back at that point. However, because it has mass

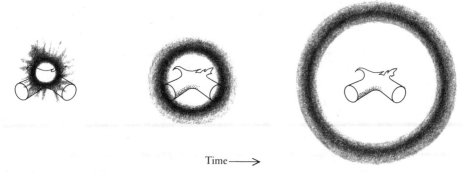

Figure 2-12 The sound wave excited by an exploding firecracker is a single expanding spherical shell of compression in the air.

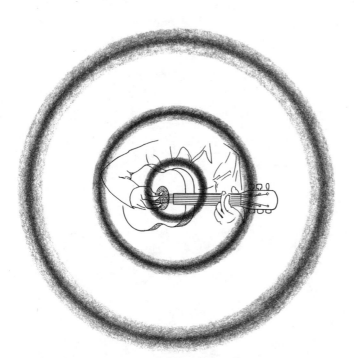

Time ⟶

Figure 2-13 A steadily vibrating object, such as a vibrating string, sends out a series of expanding shells of compression. If the object vibrates fast enough and with enough force, we can hear the succession of shells as a tone.

and, thus, inertia, it does not simply return to a straight line after it has been released, but moves on, to bend on the opposite side. Here the force of tension acts in the opposite direction and pulls the string back again. Thus the tension and mass of the string act together to keep it vibrating from side to side (see Figure 2-14).

We can look at the way in which vibrating strings move from side to side in another way: We can consider the vibrations to consist of waves that travel along the string and are reflected at its ends. If we displace and release a string, we set up a wave that travels down the string to the end, where it is reflected as a sort of echo; then it starts back along the string in the other direction as a bend on the other side (see Figure 2-15).

Under favorable conditions, waves in strings can be both seen and felt. Inside the glass façade of the New York State Theater at Lincoln Center hangs a sort of curtain made of separate strands of metal beads strung on wires or cords. During an intermission, I was standing on a walkway near the very top of these strings of beads, and I idly plucked a string to see how a wave would travel down it. A noticeable transverse wave did indeed travel down and then back up the string of beads.

I wondered if I could observe such a wave at home. First, I fastened a heavy thread to a door knob and unwound several tens of feet. Then I plucked the string. The results were inconclusive: I couldn't be sure that I

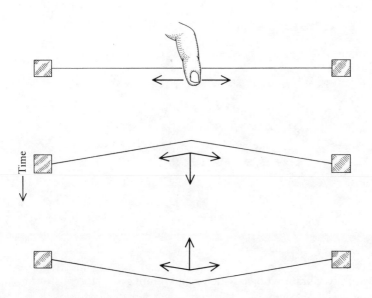

Figure 2-14 The force of tension (arrows) and the inertia of a plucked string keep the string vibrating back and forth.

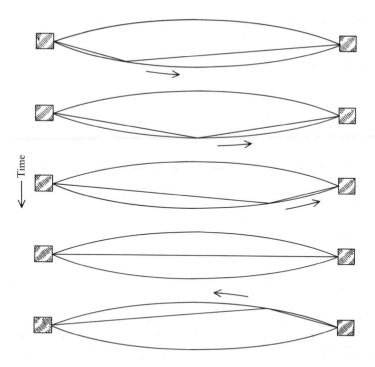

Figure 2-15 A plucked string appears to widen into a curved ribbon whose curved edges narrow to a point at each end. However, the actual shape of the string at any moment is a straight line sharply bent at one point. These "snapshots" of the bend show it traveling as a transverse wave along the stretched string. In the last three snapshots, the wave has been reflected at the right end of the string, and the bend is moving leftward on the opposite side of the string.

had observed waves, and the thread was a terrible mess when I had rewound it on the spool. I had better luck outside, with a nylon fishline. After tying one end to a tree, I unwound about sixty feet of line, stretched it lightly, and plucked the end that I held. I couldn't convince myself that I could see a wave, but I could certainly feel it as a series of jerks or pulses. The wave that I had set up traveled to the tree, was reflected there as an echo, and was reflected again at the spool that I held, traveling back and forth several times before dying out. I felt a little jerk each time the wave reached me.

Whether waves travel longitudinally in air, or transversely along stretched strings or in water, they are reflected when they encounter a solid, immovable obstacle. (The reflection of waves is discussed in Appendix E.) A reflected wave is called an *echo*. How long does it take an echo to return? That depends on the distance and on the speed of the wave. In a stretched string, the velocity, v, of the waves, measured in meters per second, is given by

$$v = \sqrt{\frac{T}{M}}$$

In the equation, T is the tension of the string, measured in newtons, and M is the mass per unit length of the string, measured in kilograms per meter. (Appendix C lists and comments on such units. Appendix D tells more about waves.) Movement of the waves is slower in a more massive string and faster in a string that is stretched more tightly.

A violin dealer once came to Max Mathews for advice about choosing strings for a violin. He knew that a steel string required more tension to produce the same pitch as a gut string, but he didn't know how much more, and he feared that trying the steel strings might damage the violin. Mathews told him to estimate the increase in tension by weighing the strings. If a steel string is more massive than a gut string, its tension must be increased in the same ratio to give the same frequency and pitch.

The velocity of sound waves is not affected by changes in air pressure, because compressed air has both greater density, which slows down the waves, and greater elasticity (resistance to compression), which speeds up the waves. However, the velocity of sound waves does increase with a rise in temperature, because the elasticity of air results from the motion of its molecules, which move faster as the temperature is increased.

At room temperature (conventionally, 20 degrees Celsius or 68 degrees Fahrenheit), the velocity of sound is 344 meters per second, 1,128 feet per second, or 769 miles per hour. This velocity was first measured, somewhat inaccurately, by Mersenne in about 1636. He measured the time interval between seeing the flash of a gun at a known distance and hearing the sound. A more satisfactory measurement was made later, in about 1750, under the direction of the Academy of Sciences in Paris. Knowing the velocity of sound in air, if we count the seconds between seeing a flash of lightning and hearing the associated thunder, we can easily tell how far away the lightning struck. The distance in feet is about 1,000 times the number of seconds — about five seconds for a mile, or three seconds for a kilometer.

Echoes, Frequency, and Pitch

Under certain circumstances, echoes demonstrate the relation between frequency and pitch. Perhaps the simplest way to hear this is to stand in front of a receding sequence of regularly spaced vertical surfaces. These could be a sequence of solid seats, as in a stadium, or the seats of a Greek-style theater (see Figure 2-16), or even a long flight of broad concrete steps. A handclap or any other sharp sound will be reflected from successively more distant surfaces, so that it returns as a series of echoes. What is the time interval between successive echoes?

Suppose that the reflecting surfaces are spaced a distance W apart. In reflection from the next farther surface, the sound has to travel an added distance W to reach the surface and an added distance W to get back (see Figure 2-17). For an observer at the sound source, the time interval T between one echo and the next is two times W divided by the velocity v of sound:

$$T = \frac{2W}{v}$$

Figure 2-16   Greek-style theater at the University of California, Berkeley.

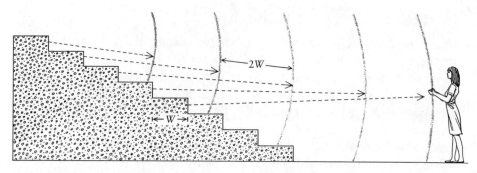

Figure 2-17 A single handclap in front of a receding sequence of steps or seats, as in a football stadium or a Greek theater, produces a periodic succession of echoes. If they are loud enough, they can be heard as a sound having a definite pitch.

The frequency f, which is the number of echoes per second, is thus

$$f = \frac{1}{T} = \frac{v}{2W}$$

If the distance W between seats is 3 feet, the frequency would be 188 echoes per second. This corresponds to a pitch of about the first G below middle C. For steps having a standard depth of 10 inches, the frequency would be 677 echoes per second, or a pitch between those of the second E and F above middle C.

It is easier to hear successive echoes from two hard, parallel surfaces. I succeeded in hearing a sequence of echoes as I stood on a concrete walkway under a projecting canopy of the Drama Building of the University of California at Santa Cruz (see Figure 2-18). When I clapped my hands, I heard a distinct sequence of claps, or pulses, of decreasing intensity. The overhang was about twenty feet above the concrete on which I stood so the reflections repeated with a frequency of about 28 per second. I didn't hear them as a pitch, but as a sequence of sharp sounds (see Figure 2-19). The hall of our house is four feet wide; for someone standing in the middle of the hall, the reflections coming alternately from *both* walls would have a frequency of

$$f = \frac{v}{w} = \frac{1128}{4}$$

Figure 2-18 The overhang at the Drama Building, University of California at Santa Cruz.

This calculation gives a frequency of 282 echoes per second — roughly, the C ♯ (C-sharp) above middle C. I can't say that I actually hear this pitch when I clap my hands in that hall, but on occasion I have heard a tonelike flutter when I clapped my hands in some rooms and halls that had hard, uninterrupted parallel walls.

Vibrations in Musical Instruments

As we have seen, with each vibration of a stretched string, a transverse wave travels the length of the string twice — once in each direction. Thus the frequency, the number of vibrations per second, is given by

$$f = \frac{v}{2L}$$

Here L is the length of the freely vibrating string, and v is the velocity with which a transverse wave travels along the string. It has already been noted that this velocity is the square root of the ratio of the tension T to the mass

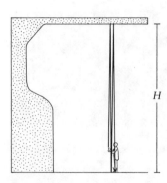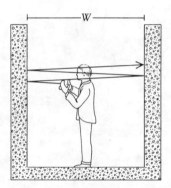

Figure 2-19 The echo from a handclap under a high overhang must travel up to the overhang and back, a distance approximately 2H, every time it returns to the observer. The echo from a handclap in the middle of a hallway must travel to the wall and back, a distance W, every time it returns.

per unit length, M. Thus, the frequency of vibration of the stretched string of a musical instrument can also be written.

$$f = \frac{\sqrt{T/M}}{2L}$$

All stringed instruments are tuned by adjusting the tension T of the strings. Thus, tightening a guitar string, for example, raises the pitch of the note that it produces, because an increase in T causes an increase in f. While playing the guitar, however, the musician continually changes the pitch of the strings by changing their length L. When a string is depressed on one side of a fret, the string does not vibrate on that side; the rest of the string, being shorter than the whole, then produces a higher note when it is plucked. Although the main change in pitch is caused by reducing the length of the string, the guitarist can also introduce minor variations into the sound of a vibrating string by wiggling the finger that holds the string down. This causes very slight fluctuations in T, and thus in f, so that the string does not produce a steady tone, but a vibrato.

Each string of any instrument must be chosen carefully so that, at a reasonable tension, the string will produce a sound of the desired pitch and loudness. In a piano, the bass strings are heavily overwound to increase their mass, besides being longer than the treble strings. Because of the great mass of the bass strings, a wave travels along them slowly; because of their long length, it has to travel far. The massive bass strings also give a loud sound. In order to make higher notes comparably loud, two strings (also overwound) are used per note in the midrange and three strings (plain) in the upper range.

In wind instruments, the frequency is determined by the time it takes a sound wave in air to travel from one end of a tube to another. A pipe of length L that is open at both ends vibrates at a frequency

$$f = \frac{v}{2L}$$

A pipe closed at one end gives a sound one octave lower than an open pipe of the same length. In an organ, for example, an open 8-foot pipe has a frequency of about 70 vibrations per second, roughly two octaves below middle C. A 16-foot open pipe or an 8-foot pipe closed at one end would sound another octave lower.

Organ pipes are arranged in ranks of pipes (also called *stops* or registers). A stop contains one pipe for each key of the keyboard and can be referred to by the length of its longest pipe. Thus all the pipes of a 16-foot stop are twice as long as the corresponding pipes of an 8-foot stop, and all their pitches are an octave lower.

The pipes of different stops differ not only in length but also in shape and composition, which gives them different *timbres*, or qualities of sound. Furthermore, two or more stops can be played simultaneously to give special timbres. *Mutation stops* are designed for just this purpose: the nazard, or $2^2/3$-foot stop, gives sounds pitched an octave and a fifth above those of an 8-foot stop; the tierce, or $1\ ^3/_5$-foot stop, gives sounds pitched two octaves and a major third above those of an 8-foot stop; the larigot, or $1^1/_3$-foot stop, gives sounds pitched two octaves and a fifth above those of an 8-foot stop. By employing mutation stops together with 8-, 4-, or 2-foot stops, organists produce strange, juicy timbres that may sound to the uninitiated like a lot of wrong notes going along with the melody.

We have seen in this chapter that the pitch of a sound depends on the frequency of the vibrations that produce it. In a siren, the frequency, and hence the pitch, of the sound depends on the number of holes that pass the nozzle each second. Echoes from a sequence of equally spaced surfaces can produce a tone whose frequency is determined by the number of echoes that reach the observer each second. In a vibrating string or a speaking organ pipe, the pitch is determined by how much time it takes a transverse wave or a sound wave to travel back and forth along the string or pipe. The next chapter explores ideas of frequency and periodicity in more detail.

A Note on Musical Pitch

Pitch originated as a musical term and has become a psychological term used to designate a perceived quality of sound.

In music, the pitch of musical sounds was perceived long before the physical basis for pitch was understood. One of the great musical (and psychological) discoveries is that for periodic musical sounds, such as those produced by the organ, strings, winds, and the human voice, pitch is tied unalterably to the periodicity or frequency with which the waveform of the sound repeats.*

Periodic musical sounds are made up of many harmonically related frequency components, or *partials*, of frequencies f_0, $2f_0$, $3f_0$, $4f_0$, and so forth. Such sounds have many perceived qualities besides pitch. One of these other qualities is shrillness, or brightness. A sound with intense high-frequency partials is bright, or shrill. A sound in which low-frequency partials predominate is not bright, but dull.

When you listen to periodic musical sounds on a stereo system, you can change the brightness by turning the tone control, but this doesn't change the pitch. The brightness depends on the relative intensities of partials of various frequencies. Turning the tone control can change the relations of the partials, but won't change the periodicity of the sound, which is the same as the *fundamental*, the frequency of the first partial, f_0.†

Sounds that are not periodic musical sounds are not as clear and distinguishable in pitch and brightness, but some of them can be granted pitch by a sort of musical courtesy. Among these are sine waves (pure tones), the tones of bells, the clucking sound that we can make with the tongue and the roof of the mouth, the somewhat related sound of the Jew's harp, and the sound of a band of noise.

Sine waves are peculiar in that they consist of a single harmonic partial. The sense of pitch that they give is not as certain as that of other periodic sounds; it can differ a little with intensity, and between the two ears. For other periodic sounds, the sense of the octave is very strong, for the partials of a sound a' (that is, an octave above sound a) are all present in sound a.‡ The sense of the octave is not strong with sine waves.

---

*There can be an ambiguity of an octave in the pitch of a periodic musical sound, because musicians sometimes (though rarely) report the pitch as being an octave away from where it falls according to its periodicity.

†The first partial need not be physically present in the sound wave, but its absence doesn't alter the pitch. This psychological phenomenon will be discussed in Chapter 6.

‡If a sound has frequency a, its partials have frequencies $2a$, $3a$, $4a$, and so forth. The octave of a has frequency $2a$, which has partials $4a$, $6a$, $8a$, and so forth. All frequencies in this latter series also occur in the first series.

Furthermore, because sine waves contain only one frequency component, their brightness is tied inextricably to their pitch.

Musically trained people react to sine waves and their pitches much as they react to periodic musical sounds. Naive people may react differently. By asking naive subjects to relate frequency changes of sine waves to a halving of pitch, psychologists found a *mel scale* of pitch (for sine waves). In the mel scale there is no simple relation between frequency and pitch; nothing like the octave shows up. I think the mel scale is a scale of brightness, not of pitch. It might be possible to check this by using musical sounds whose brightness and pitch could be varied independently.

The sounds of orchestral bells and of tuned bells are not periodic, and these sounds do not have all the properties of periodic musical sounds. One can play tunes with bells, and the pitches that are assigned to bells can be explained largely in terms of the frequencies of prominent, almost-harmonic partials.

Clucking sounds and shushing sounds (bands of noise) have a brightness, but no periodicity. Oddly, *we can play a recognizable tune with these sounds*, even though they cannot be heard as combining into chords or harmony. Apparently, in the absence of a clear pitch, brightness can suggest pitch. This seems natural. When we play a scale on a musical instrument, the brightness increases as we go up the scale. But the "pitch" of clucks or bands of noise is only a *suggestion* of pitch. It depends on the frequency at which the brightness spectrum peaks, and this (and therefore the "pitch") changes when we turn the tone control.

Even periodic sounds can be constructed to give unusual pitch effects, but such sounds are *not* produced by musical instruments. For periodic musical sounds, the pitch is tied firmly to their periodicity, the frequency of the first harmonic partial. The only mistake we can make in "confusing" pitch, a sensation, with periodicity, the numerical frequency of the fundamental, is that of offending psychologists.

3 *Sine Waves and Resonance*

We have seen that sound waves travel from the source of sound to our ears as fluctuations in air pressure. Different sorts of fluctuations cause us to hear a wonderful variety of different and interesting sounds, sounds that we can identify and appreciate. Among these are musical sounds, in which the air pressure rises and falls almost periodically with time and which have a pitch that corresponds to the frequency of this nearly periodic rise and fall in air pressure.

We can pick up the variation of sound pressure with a microphone, and we can use a cathode-ray oscilloscope to trace out the way that the sound pressure varies with time. Thus we can see the waveform of a sound wave. Is there a method that allows us to analyze sound waves and understand why the ear responds differently to different sounds? There is, indeed, and that is the subject of this chapter.

Sine Waves

Among the sounds used in the laboratory there is one, called a *pure tone*, or *sine wave*, in which the air pressure rises and falls sinusoidally with time. A mathematical explanation of sine waves is given on pages 61–63.

A sine wave is simply a tone for which the air pressure varies sinusoidally with time. It is a mathematical function that has unique and important properties. We can represent any periodic variation of air pressure over time as a sum of sinusoidal components (as we will see later). Furthermore, in responding to various sounds, the mechanism of the ear partly sorts out sinusoidal components of different ranges of frequency. When we listen to a musical sound, such as that of a voice or of a clarinet, different nerve fibers that go from the ear to the brain are excited by different ranges of the sinusoidal components. We can find these ranges by analysis of the sound wave.

The shape of a sine wave is completely described by three attributes: *amplitude, period,* and *phase.* First, the *amplitude* represents the maximum displacement of the varying quantity from its average value. In Figure 3-1, the amplitude is *h*. In a sound wave, air pressure periodically rises to a high pressure (*P*) and falls to a low pressure (−*P*), above and below the average air pressure.

Second, the *period, T,* of a sine wave is the time between amplitude peaks, usually measured in seconds. The reciprocal of *T*, 1/*T*, therefore gives the number of peaks per second, which is the *frequency, f,* of the sine wave:

$$f = 1/T$$

Frequencies used to be started as cycles per second, or cycles for short. Today the term *hertz*, honoring the physicist Heinrich Hertz (1857–1894), is used to designate cycles per second. The abbreviation of hertz is Hz.*
For example, a sine wave with a frequency of 440 Hz has a period given by

---

*Often people use *frequency* to describe the rate of oscillation of complex waveforms that are made up of many sinusoidal components of different frequencies. In this book I try to use *frequency* and *hertz* only for sine waves, and *periodicity* for the number of cycles per second of complex waveforms. This may seem awkward, but it is unambiguous.

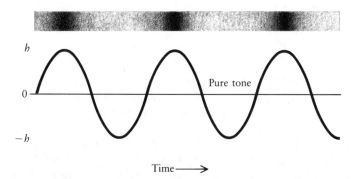

Figure 3-1   A sine wave, or pure tone.

$$T = 1/440 = .0022727 \text{ seconds}$$

Similarly, a sine wave with a period of 1/1000 or 0.001 second will have a frequency of

$$f = 1/0.001 = 1,000 \text{ Hz}$$

Third, a sine wave has a *phase*. The two sine waves shown in Figure 3-2 have the same frequency and amplitude, but have different phases because they reach their amplitude peaks and cross the horizontal axis at different times.

Why are pure tones, with their sinusoidal variation of air pressure, important for understanding musical sound? We almost never hear pure tones, except in the laboratory, or when we listen to a tuning fork that isn't struck too hard. Pure tones sound very uninteresting. When the pitch is low, they sound like the hum of a malfunctioning radio. Pure tones of higher pitch are steady but not bright or interesting whistles. Pure tones are also, in many ways, odd and unnatural sounds. In a reverberant room we can't sense the direction from which they come. Some people hear the pitch of a pure tone differently in the left and right ears; this is called *diplacusis binauralis*. Even for people with normal hearing, pure tones

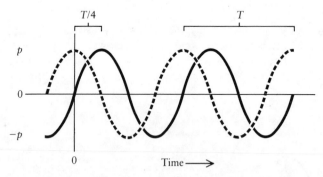

Figure 3-2 A sine wave is characterized completely by three quantities: its *amplitude* or extreme height, its *period* (the time between one peak and the next), and its *phase*, which we can take as the time when the wave crosses the axis when going upward. In this figure, the solid curve crosses the axis going upward at time $t = 0$. The dashed curve peaks at $t = 0$ and crosses the axis going upward at a time $T/4$ earlier (at $t = -T/4$, in which T is the period of the wave). The solid curve is called a *sine* curve, and the dashed curve is called a *cosine* curve. A sine curve and a cosine curve differ only in their phases.

change pitch noticeably when they are made very loud; this effect is nearly absent for musical sounds. Finally, in the low range of pitch, below middle C, pure tones aren't nearly as loud as instrumental sounds of the same power and pitch.

Happily, our ears encounter pure tones (that is, sine waves) only in acoustical laboratories, or sometimes as the output of inferior or ill-used electronic musical synthesizers. The importance of sine waves lies not in their characteristic sounds, but in their inextricable association with the systems that produce and process actual musical sounds, systems such as vibrating strings, or columns of air, or bells and gongs, or drumheads. These are, approximately at least, *linear systems*, and sine waves are the mathematical way in which the time variations of the vibrations of linear systems are expressed.

We are all familiar with stereo amplifiers. Such amplifiers are linear. If you put one frequency in (via a radio tuner or CD player for example), you get only that frequency out, not other sounds. If you double the amplitude of the input signal, you double the amplitude of the output signal. These are more than mere characteristics of linear systems; a system *is* linear if these things are true.

But if you put too intense a sine wave into the input of any actual amplifier, the sine wave will produce extraneous sounds in the output. The amplifier overloads, and you will hear frequencies in the output that weren't in the input. Further, the amplitude of the output signal is no longer directly proportional to that of the input signal. Physical systems aren't exactly linear, whether they are amplifiers or the mechanical parts of the middle and inner ear that vibrate in accord with the sounds about us. But at small amplitudes, actual physical systems do behave in a linear fashion.

This is true of strings that vibrate when struck or plucked, and of columns of air in organ pipes or horns, and of bells and gongs and drumheads. Sine waves also have a crucial relation to such vibrations. When such a resonant system is set in motion, the overall vibration is found to be made up of a host of different patterns of vibration called *modes*. The vibration of each mode varies sinusoidally with time. Or rather, the vibration functions as a decaying sine wave, a wave that loses some constant fraction of its amplitude in each short successive period of time (such decay is called *exponential*). Thus, at their very source, musical sounds are associated with sine waves — but with a collection of sine waves of many different frequencies.

The French mathematician François Marie Charles Fourier (1772–1837) invented a type of mathematical analysis by which it can be proved that any periodic wave, however it may have been generated, can be represented as the sum of sine waves having the appropriate amplitude, frequency, and phase. Furthermore, the frequencies of the component waves are related in a simple way: They are all whole-number multiples of a single frequency, for example, f_0, $2f_0$, $3f_0$, and so on.

A Fourier representation of a complicated wave can require very many components, even an infinite number. However, to approximate a wave to a desired degree of accuracy, fewer components may be adequate. A deceptively simple-looking wave whose Fourier representation requires an infinite number of components is the *square wave* (see Figure 3-3). A square wave having amplitude 1 and frequency f_0 can be represented as the sum of sine waves having frequencies f_0, $3f_0$, $5f_0$, $7f_0$ (and so on, indefinitely), amplitudes 1, 1/3, 1/5, 1/7 (and so on, indefinitely), and the proper phases (see Figure 3-4). (If the number of components being added is finite, as it will be in any real amplifier, the resulting square wave will not have exactly sharp corners; that is, it will be somewhat distorted.) We should note that square waves are peculiar in being made up only of odd frequency components. Most periodic sound waves consist of both odd and even frequency components (although closed organ pipes and some wind instruments do have predominantly odd-frequency components).

There are three different systems for naming the sinusoidal frequency components of a periodic sound, as shown in Table 3-1. Notice that the number of the harmonic or partial is the same as that of the relative frequency; for example, $5f_0$ is the fifth harmonic or the fifth partial. However, $5f_0$, the *fifth* harmonic, is the *fourth* overtone, so it is important not to confuse these three terms.

It is convenient to use the term *harmonic* to deal with strictly periodic sounds. However, some musical sounds (usually percussion) consist of frequencies that are not harmonic, that is, they are not integral multiples of the lowest frequency. For example, the frequencies of a "free" (lightly supported) vibrating rod or bar might be

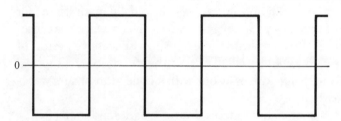

Figure 3-3 A square wave.

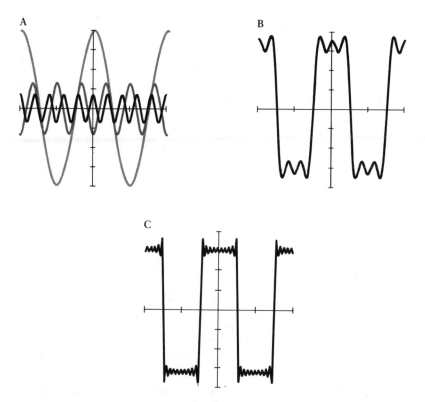

Figure 3-4   A. Three waves, of frequencies f_0, $3f_0$, and $5f_0$. B. The wave that results from adding together the three waves in part A. They roughly approximate a square wave. C. The approximation to a square wave produced by adding together the first nineteen components in the same series as in part A.

Table 3-1. Systems for Naming Frequency Components

Frequency	Harmonics	Overtones	Partials
f_0	Fundamental	Fundamental	First partial
$2f_0$	Second harmonic	First overtone	Second partial
$3f_0$	Third harmonic	Second overtone	Third partial
$4f_0$	Fourth harmonic	Third overtone	Fourth partial

Figure 3-5 The first twelve harmonies of C_2. The representations of the pitches of the seventh and eleventh harmonics are approximate.

$$f_0, 2.756f_0, 5.404f_0, 8.933f_0, \text{etc.}$$

It would sound rather illogical to call these higher frequencies "nonharmonic harmonics," so instead they are called "nonharmonic partials." It is also convenient that the numbering of the partials agrees with the numerical order of the frequency. The lowest frequency is always the first partial, the next higher frequency is the second partial, and so on.

In principle, a sum of sine waves can be used to represent any sound. If the sound is of finite duration, as actual musical sounds are, it cannot be exactly periodic over all time, as a Fourier series of harmonic partials implies. In representing exactly sounds of finite duration, we must include components of *all* frequencies by using a *Fourier integral*.

Let us also consider "noisy" sounds, such as the hiss of escaping air or the *sh* or *s* sounds of speech. Such sounds can be represented as the sum of sine waves that have slightly different frequencies. If you sound the *sh* twice, the waveforms won't look exactly alike. The power of the sound in any narrow range of frequencies will be about the same, but the amplitudes and phases of the individual frequency components won't be identical. Nevertheless, the two *sh* sounds will *sound* just the same; we will *hear* them as being the same.

Trying to represent actual sounds as sums of *true* sine waves, which persist from the infinite past to the infinite future, is a mathematical artifice. Consider the (nearly) periodic sounds produced by musical instruments. A sum of harmonically related sine waves doesn't correctly represent such a sound, because the sound starts, persists a while, and dies away.

In practice, we use the *ideas* of sine waves and their frequencies and amplitudes to characterize musical sounds, and other sounds as well. The measurements we *really* make are those suitable for our purposes, and are as accurate as they need be. They are similar in concept to Fourier series

Figure 3-6   Hermann von Helmholtz.

and Fourier integrals but aren't quite the same. Let us consider how we actually think of and analyze sounds.

Resonance

In this book *On the Sensation of Tone as a Physiological Basis for the Theory of Music* (1877), the great nineteenth-century scientist Hermann von Helmholtz (1821–1894) made extensive use of the representation of musical sounds as sums of sine waves. In his experimental work, he analyzed sounds by means of *resonators* that respond strongly to sinusoidal components near a particular *resonant frequency*. This will be our approach to the properties of musical sounds, both in analysis and synthesis.

In the nineteenth century the only available resonators were the Helmholtz resonators that Helmholtz himself devised. These were usually hollow glass spheres that had two short tubular necks, diametrically opposite one another. One opening was put to the ear, the other directed at the source of a periodic sound. If the sound contained a harmonic whose frequency was equal to or close to the resonant frequency of the cavity of the resonator, the resonator would amplify the harmonic, so that it could be heard separately. The sound in the resonator would also persist even after the source of periodic sound had suddenly been turned off. By using a succession of such resonators, Helmholtz could search out and estimate the strengths of the harmonics of a periodic sound. He could also find the

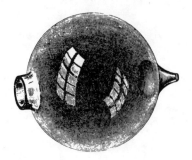

Figure 3-7   A Helmholtz resonator.

frequencies of nonharmonic partials, such as those of bells and gongs (see Chapter 13).

We can use a piano to experience something of what Helmholtz did. Hold the damper or sustaining pedal (the "loud" pedal") down (this lifts the felt dampers off the strings), and whistle near the strings. After you stop whistling, you will hear a ghostly persistence of the note that you have whistled. This dies away with time. The piano strings act as resonators. Those that can vibrate at the frequencies of the whistled note do so; those that can't, don't.

A taut piano string can vibrate at more than one frequency. The lowest, or *fundamental*, mode of oscillation was described in Chapter 2 (see Figure 2-14). These vibrations can be regarded as sine waves traveling along the string with a constant velocity v, and being reflected repeatedly at the ends of the string. If L is the length of the string, and if the wavelength (distance between peaks of the sine waves) is $2L$, we get the pattern of vibration shown in part A of Figure 3-8, in which the resonant frequency f_0 is

$$f_0 = \frac{v}{2L}$$

The next three resonant frequencies, $2f_0$, $3f_0$, $4f_0$, have the patterns or modes of vibration shown in parts B, C, and D, respectively.

Plucking or striking a stretched string excites many of these vibrations and produces a complex sound wave made up of many harmonics. The length, mass, and tension of the string determine the periodicity and pitch of the sound that is produced, but do not determine the exact waveform. The relative strengths of the various harmonics depend on whether we pluck or strike a string, and on where along its length we pluck or strike it. We can hear the difference in the sound quality or timbre. For example,

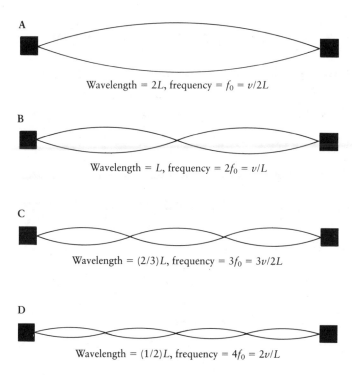

A

Wavelength $= 2L$, frequency $= f_0 = v/2L$

B

Wavelength $= L$, frequency $= 2f_0 = v/L$

C

Wavelength $= (2/3)L$, frequency $= 3f_0 = 3v/2L$

D

Wavelength $= (1/2)L$, frequency $= 4f_0 = 2v/L$

Figure 3-8 A vibrating string can have several resonant modes, which correspond to standing waves of different wavelengths and frequencies. These are the natural harmonics of the string. Parts A through D show the wavelength and frequency of the first four harmonics.

the plucked string of a harpsichord sounds quite different from the struck string of a piano. When we excite a string by bowing it, we get a persistent sound of yet another character, even though the resonances of the string may be the same.

Physical resonators are used in some musical instruments. The vertical metal tubes under the wooden bars of xylophones and marimbas, and the bamboo tubes under the brass bars of gamelan instruments, act as resonators that intensify and prolong certain of the partials generated by striking the bars, specifically, the partials that correspond to pitches of the scale. Many European baroque instruments, such as the viola d'amore, the viola bastarda, and the baryton, made use of resonators called *sympathetic strings*. These affected the relative intensities of the partials produced by bowing a string, and emitted a sustained tone even after bowing was stopped or fingering was altered.

Figure 3-9 Gamelan.

Figure 3-10 Jazz trumpeters Howard McGee (right) and Miles Davis; McGee's embouchure is relaxed; he is playing a relatively low note.

The long tubes of brass instruments resonate at harmonically related frequencies. Which frequency is excited when the instrument is played depends on how the player constricts his or her lips. The tubular structures of woodwind instruments are resonators whose resonant frequencies are controlled by opening or closing various holes.

Resonances also have a noticeable effect on the timbre of musical sounds. The vocal tract has several resonances that emphasize various ranges of frequency in the sound produced by the vibration of the vocal folds. By changing the shape of the vocal tract, we change the frequencies of these resonances, or *formants*, which determine what vowel sound we produce. Table 3-2 gives the first three resonance or formant frequencies for some vowel sounds. The resonances of the soundboard of the violin greatly affect its timbre. Figure 3-12 shows, as a function of frequency, the ratio of the sound radiated by a good violin to the motion of the bridge caused by bowing the string. The rising and falling curve reduces the intensities of some partials and increases those of others. The suppression of some partials is important for the musical quality of the violin tone. In poor violins, the sound is harsh; a curve like that shown here does not rise and fall so markedly.

Conventional musical instruments use mechanical or acoustical resonators in producing musical sounds. When analyzing a sound, we pick the

Figure 3-11 Jazz trumpeter Woody Shaw: his tense embouchure shows that he is playing a relatively high note.

Table 3-2. The First Three Formants of Selected Vowels

Formant number	Cycles per Second per Vowel							
	Heed	Hid	Head	Had	Hod	Hawd	Hood	Who'd
1	200	400	600	800	700	400	300	200
2	2,300	2,100	1,900	1,800	1,200	1,000	800	800
3	3,200	2,700	2,600	2,400	2,300	2,200	2,100	2,050

sound up with a microphone and determine its frequency content electronically. Early analyzing devices made use of electric circuits called *filters*. A filter can be described by a curve that shows how the ratio of output amplitude to input amplitude varies with frequency (see Figure 3-13).

Present-day devices for frequency analysis make use of digital techniques. In a digital *spectrum analyzer*, the Fourier spectra of short sections of the sound wave are computed by a specialized digital computer and displayed on a televisionlike screen. Or the analysis can be made by a computer with a sound-processing board. The curve on the screen shows intensity as height, and frequency as distance to the right (as shown in Figure 3-14). The curve in this figure has one peak or *line*, representing a

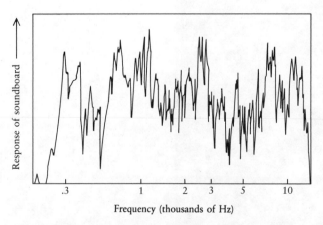

Frequency (thousands of Hz)

Figure 3-12 The body of a violin resonates more at some frequencies than at others. This is a plot of the intensity of the sound wave produced by exciting the bridge of a famous Guarneri violin with a sinusoidal source of constant amplitude and increasing frequency.

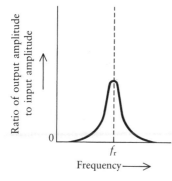

Figure 3-13 A resonator or electronic filter responds only to a narrow range of frequencies. In this curve the ratio of output amplitude to input amplitude is plotted against the frequency of the input wave. The output peaks the resonant frequency, f_r.

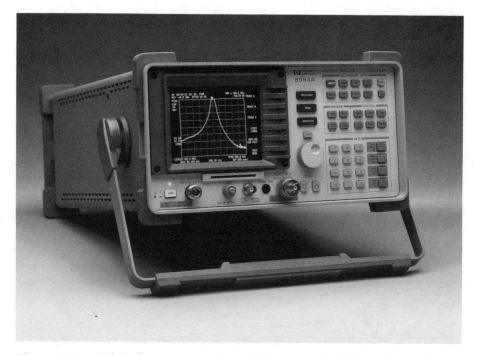

Figure 3-14 A digital spectrum analyzer displays a graph of intensity against frequency on a cathode-ray tube. (*Photo of HP 8594A portable spectrum analyzer courtesy of Hewlett-Packard, Signal Analysis Division, Rohnert Park, California*)

single frequency. The line spectrum of a musical sound has many sharp peaks, each corresponding to a single partial or frequency component.

In a musical sound, the intensity of the various partials can change with time. This can be depicted by a perspective drawing, as in Figure 3-15, which shows a sound having six harmonic partials. With the onset of the sound, the intensity of each partial rises to a peak; as the sound dies away, the intensity falls. The higher partials have a lower peak intensity than do the lower partials. As will be seen, such a representation isn't mathematically accurate, because an actual sine wave can't change intensity with time; it goes on at the same intensity forever.

The perspective drawing in Figure 3-16 is made up of a sequence of line spectra of a tom-tom. We see that the many frequency peaks are not truly separate, as in an ideal line spectrum.

We can understand why we cannot represent both frequency and time precisely if we consider another sound-analyzing device called a *sonograph* or *sound spectrograph*, which produces sonograms. Frequency is displayed vertically, and time is displayed horizontally; intensity is indicated by the darkness of the shading. The sonograms in Figure 3-17 are those of a human voice.

The broad horizontal bands in both sonograms represent the resonant frequencies of the vocal tract, or formants. Such resonances change rela-

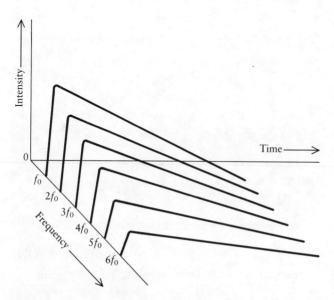

Figure 3-15 A perspective drawing of a sound in which the intensities of the six harmonics shown rise and fall with time.

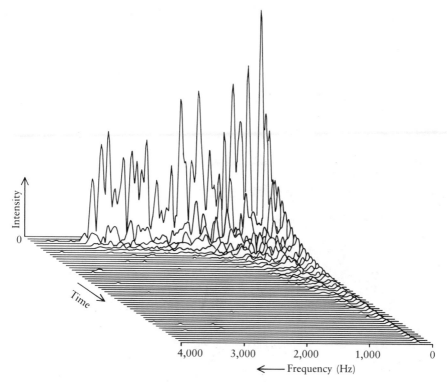

Figure 3-16 Depiction of how the spectrum of the sound of a tom-tom changes with time after the tom-tom is struck.

tively slowly with time as we speak. However, the upper and lower sonograms differ in detail. In the upper sonogram we see horizontal striations. They represent the individual harmonic partials of the voice, and their frequency separation is equal to the pitch frequency. In the lower sonogram we see vertical striations, separated by a time equal to the period, T, of the vibration of the vocal folds (the pitch frequency is $1/T$); but we don't see any horizontal striations. The upper sonogram was made with a narrow-band filter, which can sort out individual harmonics. However, such a filter cannot respond to rapid changes of sound pressure, so we don't see vertical striations corresponding to individual pitch periods. The lower sonogram was made with a broader-band filter, which responds to several harmonics at once; hence we do not see horizontal striations corresponding to individual harmonics, but we do see vertical striations corresponding to individual pitch periods.

Using filters to analyze sound means that, if we depict frequency in fine detail, we can't depict time in fine detail, or, if we depict time in fine

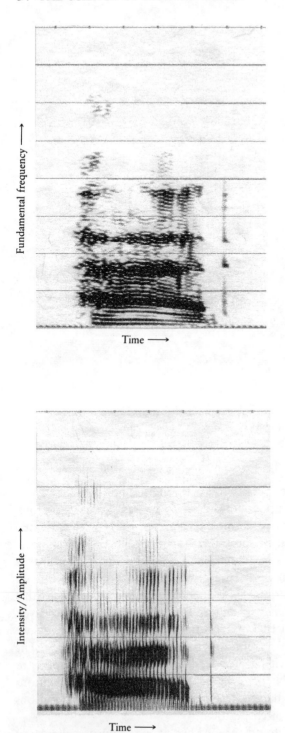

Figure 3-17 Sonograms of a human voice, saying the vowel *a* in "had."

detail, we can't depict frequency in fine detail. This is why Figure 3-15 is only qualitatively (not quantitatively) correct, for it seems to show both frequency and time very accurately.

Sonograms are usually used to get an approximate idea of how amplitude or power varies with frequency and time. However, if we take successive spectra of properly overlapping portions of a waveform in the right way we can reconstruct the waveform from the successive spectra. This is easier if in taking the successive spectra we compute the phases of the sinusoidal components as well as their amplitudes.

A device that is built or programmed for such analysis into successive spectra and resynthesis into the original waveform is called a *phase vocoder*. Besides being used in studying musical sounds, phase vocoders are used to alter sounds in useful ways by modifying the spectral representation before resynthesis. Phase vocoders can be used to shift the pitch of sounds, or to stretch them out or compress them in time, or both at once.

Although no musical sound is truly periodic in persisting unchanged forever, most musical sounds are nearly periodic. They can be approximated very closely by a fairly small number of sinelike waves whose amplitudes rise and fall slowly with time, and whose frequencies are nearly harmonically related; so they can be represented by a succession of changing line spectra. The vibration of a piano string dies away slowly; in its line spectrum, the peaks that represent the various harmonics decrease in height as the intensity of the sound decreases. Bells and gongs (see Chapter 13) do not produce periodic sounds, but they do have line spectra that decrease in amplitude gradually as the sound dies away.

Suppose we measure the spectrum of a sound and get a smooth curve rather than a series of spikes, as in Figure 3-18. Such a sound can be

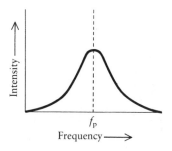

Figure 3-18 Spectrogram of a noise that is made up of an infinite number of sinusoidal frequency components whose phases are random and therefore don't peak at the same time. Instead, they give the bell-curve distribution familiar from statistics. A noise with a spectrum like this one will give a sense of pitch at the frequency f_p, at which the spectrogram peaks.

represented, not by a finite number of frequency components, but by a continuous distribution of partials throughout a range of frequencies. You can make such a sound by whispering a vowel. The frequency (or frequencies) at which the spectrum peaks depends on the resonance (or resonances) of the vocal tract. If you whisper an *e*, you will get a "high-pitched" sound; if you whisper a *u*, you will get a "low-pitched" sound. This is because the resonances of the vocal tract occur at higher frequencies for an *e* than for a *u*. In a whispered *e*, the resonances emphasize high-frequency components in the breathy sound. In a whispered *u*, the resonances emphasize lower-frequency components.

Noise and Pitch

Noise can be narrow-band or broad-band, that is, the peak, or hump, in the spectrum can be narrow or broad. If the spectrum has no peak, so that all frequencies within a range are present equally, the noise is called *white noise*. This is the even, breathy, faintly frying sound that you get on some noisy telephone connections, or when you turn the volume control of an AM radio up when it is not tuned to a station.

If we gradually narrow the width of a spectral peak such as that shown in Figure 3-18, we will get a more and more pronounced sense of pitch, which corresponds to the frequency (f_p) of the peak. When the peak becomes very narrow, the noise ceases to sound noisy. Instead, it sounds like a sine wave that wavers randomly in amplitude, but only a little in frequency (see Figure 3-19). Max Mathews (in an early computer piece called *The Second Law*) and others have produced computer music that uses noise of various bandwidths, both narrow-band tonelike noise and broader-band shushing noise.

Sounds that have a continuous spectrum are not necessarily noises. If the phases of all the sinusoidal frequency components are equal, so that

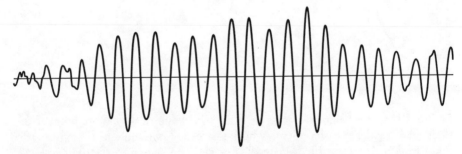

Figure 3-19 The waveform of a narrow band of noise, a sinelike wave whose amplitude varies randomly with time, but whose frequency is nearly constant.

they are peak at the same instant, or if the phases change slowly and smoothly as the frequency changes, we get a single pulse rather than a chaotic, persistent noise.

Part A of Figure 3-20 shows a short pulse or burst of pressure that would sound like a click, and part B shows that the sinusoidal components of which it is composed span a broad range of frequencies, from zero on up. Part C shows a pulse that has a broader rise and fall of amplitude. It will be more thumplike than clicklike, and part D shows that its spectrum spans a narrower range of frequencies. In general, the shorter the pulse of sound, the broader the range of frequencies. Roughly, the width of the frequency range or band (B) is inversely proportional to the duration (D) of the sound in seconds:

$$B = 1/D$$

Most short sounds aren't simple rises and falls in pressure like those in this figure. When we strike a block of wood, we get a sort of click, but the

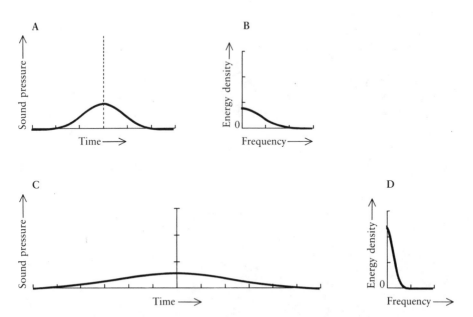

Figure 3-20 The spectrum of a single pulse of air pressure is not a line spectrum, but a continuous spectrum made up of components of a continuous range of frequencies. If the pulse is short (part **A**), the spectrum is broad (part **B**), and the pulse sounds like a click. If the pulse is longer (part **C**), the spectrum is narrower (part **D**), and the pulse sounds like a thump.

pressure tends to oscillate up and down for a short time. Similarly, when we strike a bass drum, we get a sort of thump, but the drumhead oscillates for a short time. The bandwidth, or range of frequencies, of such sounds is still proportional to the duration of the sound, but the peak of the spectrum is at the frequency with which the struck object oscillates: the larger the object, the lower this frequency of oscillation, and the duller the sound. For example, in part A of Figure 3-21, we have a short section of a waveform, two cycles long, and in part C a longer section, four cycles long, twice the duration. Such short sections of sine waves give a sense of whether the sound is "dull" or "bright." In parts B and D, the peak of the spectrum is at the frequency of f_0 of the sine wave, but the spectrum of the longer, four-cycle section is half as wide as the spectrum of the short, two-cycle section.

When a sine wave is turned off or on abruptly, as shown in Figure 3-21, this excites resonant devices of many frequencies. Indeed, we hear a click when the wave is turned on or off abruptly. Anyone who listens to the output of an audio oscillator, and makes and breaks the connection

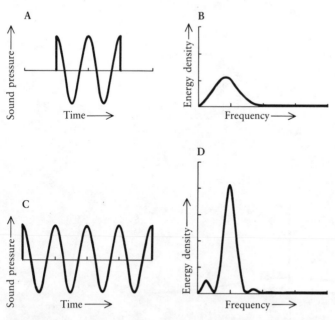

Figure 3-21   A short section of a sine wave of frequency f_0, as in part **A**, has a broad spectrum centered on the frequency f_0, as in part **B**. A longer section of the same sine wave, as in part **C**, has a narrower spectrum, as in part **D**, still centered on the frequency f_0. When the sine wave is turned on abruptly, as shown in parts **A** and **C**, we hear a click.

with the speaker by flipping a switch, will hear such a click. We can get the effect of hearing a short pure tone or sine wave without a click by increasing and decreasing the intensity gradually, as shown in Figure 3-22. If the slowly rising and falling sine wave has a total of about 40 or more complete cycles, it sounds like a short burst of tone without a click. If it has 4 cycles or fewer it sounds like a click without a clear pitch, though, like a noise, it can sound "high" or "low," or "bright" or "dull," depending on the frequency of the sine wave. For intermediate numbers of cycles, we hear both a click and a pitch. The click part of the sensation becomes more prominent as the number of cycles is reduced, and the sensation of clear pitch fades into just "high" or "low," or "bright" or "dull." The appearance of the click occurs at a somewhat smaller number of cycles for sounds heard in a reverberant room than for sounds heard through headphones.

The practical result here is that we can add "slowly changing sine waves" to produce periodic, pitched sounds of finite duration. Mathematically, such a "slowly changing sine wave" is not a sine wave at all, but it approximates many of the properties of a sine wave. Both instruments for measuring and the human ear respond to such a "slowly changing sine wave" as if it were a pure tone that changes slowly in amplitude, frequency, or phase. However, if we change sine waves too fast, we will hear a click or the twang of a plucked string.

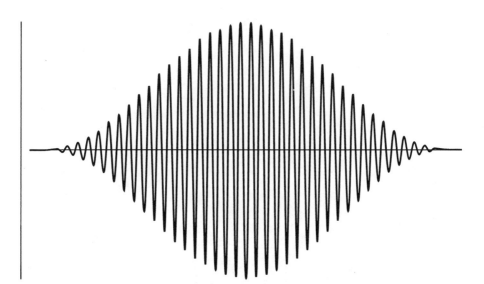

Figure 3-22 If a sine wave is turned on and off slowly, we don't hear a click. (Mathematically, this is not a sine wave but, for practical purposes, it sounds like one.)

On the screen of a spectrum analyzer, a pure tone of slowly increasing amplitude appears as a line or peak of increasing height, and the corresponding sound grows louder and louder without changing pitch (much). If we change the frequency of a sine wave gradually, the line on the spectrum-analyzer display moves to the left or right, and we hear a fall or rise in pitch. A periodic change in frequency (a *vibrato*) is heard as a periodic change in pitch if it is slow, and the line on the spectrum-analyzer display jiggles back and forth. Above a vibrato rate of about six per second, we can no longer hear vibrato as a change in frequency; rather, vibrato gives an entirely new and pleasant quality to a musical sound.

All that I have said about a single slowly changing sine wave also applies to sums of slowly changing sine waves. Musical tones which are combinations of harmonically related sine waves, each of which changes slowly in amplitude or frequency, are heard as tones of changing loudness or pitch. In the chapters that follow, many sounds will be described as combinations of sine waves, or partials, even though the amplitudes and frequencies of the partials change with time.

A Note on Sine Waves

A sine wave is not just any wiggly curve that rises and falls with time; it is a precise mathematical function that can be described very simply. The crank illustrated in part A of Figure 3-23 turns with a constant speed, making one revolution every T seconds. The height of the crank at any particular time is shown as h. We can draw a sine wave by plotting the height of the crank against time.

To begin to do this, as shown in part B of the figure, place equally spaced points around the circumference of a circle. The figure shows eight points, numbered from 0 to 7. For each successive point, starting with 0, measure the height of the point above or below the horizontal axis line through the center of the circle. Plot the successive heights of points 0, 1, 2, 3, and so forth, as in part C, at successive equally-spaced distances along the horizontal line; then connect these points with a smooth curve to draw the sine wave. If you use more equally-spaced points around the circle, you can draw the sine wave more accurately, as in part D.

Mathematically, the relation between h (or amplitude) and time t is written $h = \sin wt$, in which w is a constant, equal to 2π times the frequency f, and $f = 1/T$, T being the period. The higher the frequency (that is, the shorter the period), the more times the sine wave goes up and down each second.

There is another type of sine wave, the *cosine*, which is simply a sine wave with a different phase. The cosine of wt is written $h = \cos wt$. The cosine reaches its peak a quarter of a period ($T/4$) before the sine does. The cosine is traced out by the height of the dashed crank in part A of Figure 3-23, which is at right angles to the solid crank, and is shown as the dashed curve in part C. To draw the cosine we simply start at point 2 instead of point 0, and plot successively the heights of points 2, 3, 4, 5, 6, and so forth.

This simple "mechanical" picture can tell us all we need to know about sine waves. Those who like mathematical puzzles can work out all sorts of interesting, useful, and important relations, such as

$$(\sin wt)^2 + (\cos wt)^2 = 1$$

A particularly interesting and important relation is

$$(\sin pt)(\sin wt) = (1/2) \left[\cos (w - p)t - \cos(w + p)t \right]$$

Here we can think of the sine wave $\sin pt$ as controlling the intensity or amplitude of the sine wave $\sin wt$. What can we discover from this simple

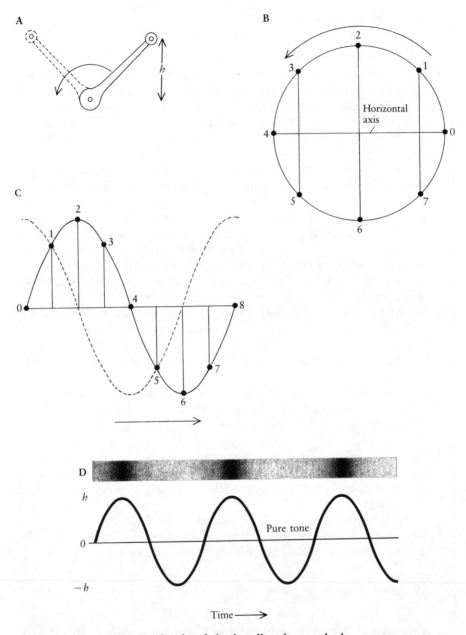

Figure 3-23   A. The height, *h*, of the handle of a crank that rotates at a constant speed varies sinusoidally with time. **B.** Equally spaced points around the circumference of a circle. **C.** When the points in part **B** are translated to a graph in which the steady passage of time is represented as movement to the right, they outline a sine wave. **D.** A smooth sine wave would be produced if every point on the circle were translated to the graph.

relation? First, we see that the product of two sine waves is simply two sine waves (since cosine waves are just sine waves that peak at different times, that is, have different *phases*). The average value of the product of two sine waves is zero, because the average value of each of the two sine waves is zero. If the sine waves have exactly the same frequency, one of the two components on the right is the cosine of zero, which is equal to one. In mathematically representing a function of time as a sum of sine waves, we make use of this property.

We can deduce another important result from this last formula. Imagine that p is very small compared with w, say, $p = 1$ and $w = 100$. Then the factor $\sin pt$ will cause the amplitude of the oscillation $\sin wt$ to vary slowly with time. Our formula tells us that

$$(\sin t)\,(\sin 100f) = (1/2)\,(\cos 99t) - (1/2)\,(\cos 101t)$$

That is, a slowly varying sine wave doesn't have only one frequency; mathematically it is made up of two or more frequencies, but these frequency components are very near the nominal frequency of the rapidly varying sine wave. This is very important, for when we turn a sine wave on or off, we still think of it as having the same frequency that it would if it persisted forever. In practice, to our sense of hearing, this is true *as long as we turn the sine wave on and off slowly enough* to avoid getting a click.

4 Scales and Beats

*A*t one time or another many of us have heard a piano tuner working on a piano. We have heard him or her strike an octave and then use a tuning wrench (or key) to tighten or loosen a string. How does a tuner know when the octave is tuned just right? The slight difference, the *beat* between the two notes disappears. What is this beat that guides the tuner?

Figure 4-1 shows the sum of two sine waves of slightly different frequencies (parts A and B). This sum looks like a single sine wave whose amplitude increases and decreases slowly with time (part C), and that is what it sounds like. The two sinusoidal components seem to merge into a single throbbing or beating sound (part D). If we make the sine waves more and more nearly equal in frequency, the beat between them becomes slower and slower. When the sine waves have equal frequencies, the beat disappears, and we hear a tone of constant amplitude.

What beat does the piano tuner listen for in tuning an octave? A single note on a piano has many harmonics, and its second harmonic is an octave higher than its fundamental tone. In tuning an octave, the piano tuner listens for beats between pairs of harmonics, as, the second harmonic of the lower tone and the fundamental of the upper tone, or the fourth harmonic of the lower tone and the second harmonic of the upper tone. For true harmonics, all such beats slow to zero for the same tuning.

Actually, piano strings have a little stiffness, which adds to the effect of tension in keeping the strings straight. As a result, the higher partials have frequencies that are a little greater than integer multiples of the

A

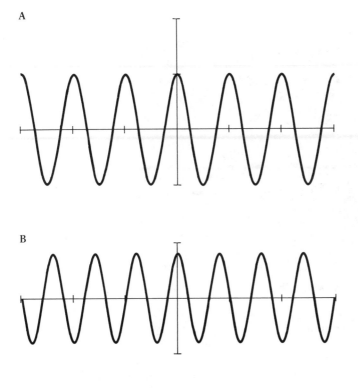

B

C

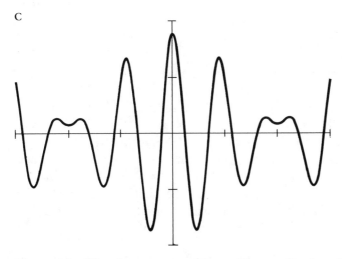

Figure 4-1 The phenomenon of beats. The amplitudes of the two sine waves in parts **A** and **B**, whose frequencies are very close to one another, add up to give the wave in part **C**, which looks very much like a sine wave of slowly varying amplitude.

Figure continued on following page.

D

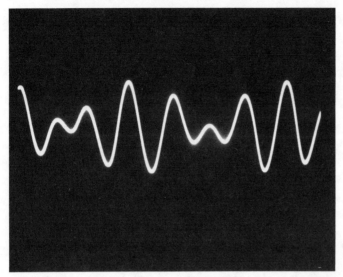

Figure 4-1 (*Continued*) The photograph of an oscilloscope screen in part **D** shows the beat between tones of 4,000 and 4,100 Hz. When the frequency difference is small, it sounds like a sine wave or pure tone whose intensity rises and falls slowly with time. As the frequencies of the sine waves are made more and more nearly equal, the rate at which the amplitude rises and falls (the frequency of the beat, or the beat frequency) gets slower and slower until, when the frequencies are equal, the beat disappears.

fundamental frequency f_0; that is, they are slightly larger than $2f_0$, $3f_0$, $4f_0$, $5f_0$, and so forth.

When octaves in a piano are tuned by the method of eliminating beats, the octaves will therefore be *stretched* a very little; that is, they will have frequency ratios slightly greater than 2. Pianos *are* often tuned with slightly stretched octaves, sometimes because of the stiffness of the strings, sometimes because the pianist prefers the brighter timbre that results. In this discussion, we will ignore the stiffness of piano strings and other practical realities, and assume that *all* musical tones have partials that are exactly harmonic, that is, that are integer multiples of the frequency of the first partial.

Frequency Ratios

Intervals other than octaves can also be tuned by means of beats. To see how, consider the piano keyboard shown in Figure 4-2. We go up a *semitone* when we go from any key (white or black) to the next key (black

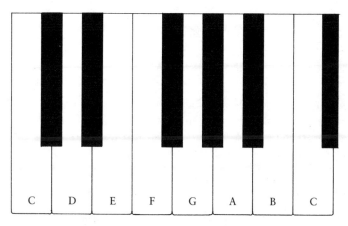

Figure 4-2 The white keys of the piano give the seven notes of the C-major diatonic scale.

or white). Certain intervals are called *consonant intervals*, because musicians and listeners consider pitches separated by these intervals to be pleasing when sounded together. These intervals are given in Table 4-1, which shows both the ideal frequency ratio between the two notes and the number of semitones of difference between them on the piano keyboard.

Figure 4-3 shows the harmonics of a C with a frequency f_0 and the G above it, which has a frequency $(3/2)f_0$. We see that the third harmonic of C has the same frequency, $3f_0$, as the second harmonic of G. If the C and G

Table 4-1 Consonant Intervals.

Name of interval	Notes (in Key of C Major)	Ideal Frequency Ratio	Number of Semitones
Octave	C-C	2	12
Fifth	C-G	3/2	7
Fourth	C-F	4/3	5
Major third	C-E	5/4	4
Minor third	E-G	6/5	3
Major sixth	C-A	5/3	9
Minor sixth	E-C	8/5	8

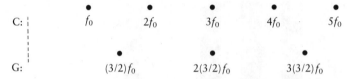

Figure 4-3 The frequencies of two pitches (such as C and G) that are an interval of a fifth apart are always in the ratio of 100:150. Their harmonics will be in the same ratio; so the second, fourth, and so forth, harmonics of G will coincide with the third, sixth, and so forth, harmonics of C.

are a little mistuned, these harmonics will produce an audible beat when the two notes are struck together. By tuning G so that it doesn't beat with C, we can assure that the fundamental frequency of G is just 3/2 that of C, in musical terms a perfect fifth.

A little thought will show that other pitches can be tuned in this way as well. A fourth has a frequency ratio of 4/3, so the third harmonic of F should have the same frequency as the fourth harmonic of C (which is C″, two octaves above the first C).* A major third has a frequency ratio of 5/4, so the fourth harmonic of E should have the same frequency as the fifth harmonic of C. A minor third has a frequency ratio 6/5, so the fifth harmonic of E should have the same frequency as the sixth harmonic of C (a G).

A friend of mine once tried to tune his piano according to such relations, thinking that by later adjustments he could get to the *equal-tempered* scale, to which a piano is really tuned. He underestimated the difficulty. Piano tuners have a systematic method of tuning, in which they tune the intervals within a single octave to have prescribed numbers of beats per second. Once they have tuned these twelve notes, they tune all the other notes in octaves above and below by the method of octaves, with no beats. This isn't what my friend did.

Here we come to the dilemma of the diatonic, or major, scale, represented by the white keys of the piano. Is there a *sensible* explanation for this scale, which is used in so many cultures? There *is* an explanation, but you must judge for yourself how sensible it is.

---

*The primes are used to flag *relative* positions of notes. As we start up, we have the notes C, D, E, F, G, A, B; in the next octave up the notes are C′, D′, E′, F′, G′, A′, B′; in the next, C″, D″, and so forth. This convention helps us keep track of *which* C we are talking about as we jump up and down the octaves in a discussion of harmonics, say. This designation is *relative*, not absolute; any note we choose can be the starting point for the primes in a given discussion.

First, we should observe that notes or tones an octave apart sound very similar. In primitive cultures, men and women sing in octaves without realizing that they are singing different notes. With some timbres, it is not uncommon to make an error of an octave in judging pitch. The psychologist Roger Shepard has likened the similarity of octaves to points on an ascending spiral. As shown in Figure 4-4, the notes C, D, E, F, G, A, B repeat again and again around the spiral as we go up the scale. But C′ an octave up always lies close (in perception) to C an octave below, and so it is for all the other notes of the scale.

With this in mind, consider Figure 4-5, which shows notes in the bass and treble clefs. In part I, we see the first six harmonics of the C an octave below middle C.* They are, successively, C, (middle) C′, G′, C″, E″, and G″. The intervals between these successive notes are consonant intervals. They are the octave (C to C′), the fifth (C′ to G′), the fourth (G′ to C″), the major third (C″ to E″), and the minor third (E″ to G″).

So far we have only three of the seven pitches of the C-major diatonic scale, C (the *tonic*), E (the *mediant*), and G (the *dominant*). A plausible way in which to get one more note is to drop down a fifth from C″, as shown in part II of the figure. This gives us F′ (the *subdominant*); the frequency ratio of F′ to C′ is 4/3.

In going from C″ to E″ to G″, as in part I, we go up first by a major third (a frequency ratio of 5/4), then by a minor third (a frequency ratio of 6/5), and so arrive at an overall frequency ratio between G and C of 3/2.

---

*It is called "middle C" because it is written on the line in the middle, between the treble and bass clefs.

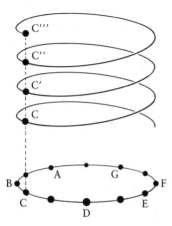

Figure 4-4 The notes of the diatonic scale as points along an ascending spiral. Here all Cs are close together and hence alike; all Ds are close together and hence, alike; and so on.

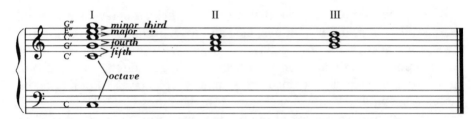

Figure 4-5 The successive harmonics of a note C give us the notes C', G', C", E", and G", and include the intervals of the octave, the fourth, the major third, and the minor third. In any octave, if we go up a major third from G', we get B', and if we go up a minor third from B, we get D". If we go down from C" by a fifth we get F', and if we go up a major third and then a minor third from F' we get, successively, A' and C". We have now derived all the pitches of the diatonic scale. In the notation shown, C' is middle C.

The fourth, or subdominant (F in the C major scale), plays an important part in Western music. It seems reasonable to go up from F' in the same pattern, first by a major third, then by a minor third, as in part II, thus basing a *major triad* on F'. This gives us A' (the *submedian*) and brings us back to C". Here A' is a minor third below C", that is $(6/5)$ A' $=$ C" and, because C" $=$ 2C', the frequency ratio of A' to C' is $5/3$.

The fifth, or dominant (G in the C major scale), plays an even more important role in Western music. Let us also base a major triad on G', again going up by a major third, then by a minor third, as shown in part III. This gives us the seventh, or *leading tone*, B', and the *supertonic*, D".

We now have all seven pitches of the diatonic scale in *just intonation* in the key of C; C, being the first note of the scale, is the *tonic*.

Equal-Tempered Tuning

Musical intervals with frequency ratios given by the ratio of integers, such as 3/2 (fifth) and 5/4 (major third), are the basis of the diatonic scale. Such musical intervals are, of course, important in music itself. Let us concentrate on the black notes in the short passage of music shown in Figure 4-6. They are C, A, D', G, C. Let us designate the frequency of C as f_0. In going up to A, we go to a frequency of $(5/3) f_0$. The interval from A up to D' is a fourth, with a frequency ratio of 4/3; so when we arrive at D', the frequency should be 4/3 that of A, which is 5/3 times f_0, or $(4.3) (5/3)$ $f_0 = (20/9) f_0$. We now go down a fifth to G, which should give us a frequency of $(2/3) (20/9) f_0 = (40/27) f_0$. We now go down another fifth to C, which should give us $(2/3) (40/27) f_0 = (80/81) f_0$. But we started

Figure 4-6 Consider the black notes in this series of chords. They proceed from C to A to D′ and back down to G to C, or by intervals of a sixth up, a fourth up, a fifth down, and a fifth down. If these were perfect intervals, in which the frequency ratio of the sixth is 5/3, of the fourth is 4/3, and of the fifth is 3/2, we would not get back to the same pitch from which we started.

out by setting the frequency of C equal to f_0, and we haven't gotten back there!

This short passage of music shows that no fixed scale of pitches can accommodate all upward and downward motions of pitch by ideal musical intervals, that is, musical intervals with frequency ratios given by the ratios of small integers. A number of failures to return to the initial pitch after such sequences of ideal intervals have been noted and named. They are shown in Table 4-2.

The comma of Pythagoras is of particular interest. If we go up repeatedly by the interval of a perfect fifth ($3:2 = 1.5$) we will reach octaves of

Table 4-2 Final "Error" in Sequences of Ideal Intervals That *Almost* return to Initial Pitch.

Name	Ratio of Difference	Difference in Cents	Motions of Pitch
Comma of Didymos, or synotic comma	$\dot{8}1/80$ $\left(\frac{3}{2}\right)^4 = \frac{81}{16} \cdot \frac{1}{4} \cdot \frac{1}{2} \cdot \frac{4}{5} = \frac{81}{80}$	21.5 cents	Up four fifths, down two octaves, down one major third
Schisma	32,805/32,768	2.0 cents	Up eight fifths, up one major third, down five octaves
Comma of Pythagoras	521,441/524,288	23.5 cents	Up twelve fifths, down seven octaves
Diesis	128/125	41.0 cents	Up one octave, down three major thirds

all the white and black keys of the piano keyboard—almost. If we go up repeatedly in a ratio close to that of a perfect fifth, a ratio approximately 1.4983, we do reach the octaves of all black and white keys, and we can get the pitches of all other keys by subsequently going down or up by octaves. This may be thought of as the source of equal-tempered tuning.

In a very real sense, perfect intervals, whose frequency ratios are the ratios of small integers, are the very foundation of music. These intervals derive from the harmonics present in musical tones. They are important to the ear. Yet they cannot be exactly realized in all the intervals of any one scale, any one system of tuning. The history of Western music is also the history of the attempts to resolve these discrepancies.

In discussing departures of intervals of a scale from ideal intervals, the *cent* is very useful. The cent is a hundredth of a semitone, or $1/1,200$ of an octave. The frequency ratio of the cent is the $1/1,200$ power (the 1,200 root) of 2, or 1.00057779. If we multiply 1,200 of this decimal number together (that is, if we raise it to the 1,200 power), we get 2, the frequency ratio of the octave. If we raise the number to the 100 power, we get the frequency ratio of the equal-temperament semitone, 1.059463.

Table 4-3 shows the intervals of the diatonic scale in cents for equal-tempered tuning, for *Pythagorean* tuning (which is designed to give perfect fifths), and for *just* tuning. Table 4-4 shows the "error" in these compromise tunings relative to the ideal ratios.

Some musicians love just temperament dearly. For example, Harry Partch built many strange instruments whose sounds were as wonderful as their names, and which were often of somewhat uncertain pitch. He had a

Table 4-3 The Intervals of Differently Tempered Scales, Measured in Cents.

Note	Equal-tempered	Pythagorean	Just
C	0	0	0
D	200	204	204
E	400	408	386
F	500	498	498
G	700	702	702
A	900	906	884
B	1,100	1,100	1,088
C	1,200	1,200	1,200

Table 4-4 Errors in Tempered Scales Relative to Ideal Ratios.

Interval	Name	Equal-tempered Error	Pythagorean Error	Just Error
C–E	Major third	+15	+22	0
D–F	Minor third	−16	−22	−22
C–F	Fourth	+2	0	0
C–G	Fifth	−2	0	0
C–A	Sixth	+16	+22	0
D–A	Fifth	−2	0	−22

harmonium justly tuned in the key of C. It sounded excellent in C, but dreadful when played in any other key.

Let us consider just temperament closely. Figure 4-6 shows that in just temperament the musical interval of a fifth from D to A is in error by 22 cents; that is why we are able to get back to C in just temperament, but at a considerable cost. This 22-center error is 6 cents larger than any error in equal temperament.

Furthermore, in equal temperament we have dealt very simply with the problem of sharps and flats. Ideally, in the interval of the minor sixth, A♭ (A-flat) should have a frequency ratio of 1.6 to be a perfect major third from the C above it. In order to make the interval E–G♯ (G-sharp) a just major third, G♯ should have a frequency ratio $25/16 = 1.5625$. In our equal-tempered scale, both A♭ and G♯ are represented by the same black key on the piano, and by a frequency ratio of 1.5874.

There are flats and sharps in diatonic scales in which the tonic or first note is not C. A fifth above C is G, the first note of a closely related key, G major. G major has six notes in common with C major. (The key of G major employs F♯ instead of F.) A fifth above G is D. The scale starting on D has two sharps, C♯ and F♯, and has only five notes in common with the C scale.

Going up through a cycle of successive fifths on the piano, we reach all keys, white and black, and on each we can build a scale. Or we can start down from C by an interval of a fifth, to reach F. The scale in which F is the first note, or tonic, has six notes in common with the C scale; in F, B♭ is used instead of B.

Johann Sebastian Bach was one of the first strong advocates of a tuning such that the intervals, though slightly "in error," are tolerable in all keys. Bach's *Well-Tempered Clavier* includes pieces in all keys. With equal-tempered tuning, all are equally in tune (or out of tune).

It is often argued that equal temperament can be offensive to musicians with keen ears. During the nineteenth century instruments with complex, almost unplayable keyboards were devised in order to allow the production of close approximations to ideal intervals; some are discussed in Helmholtz's book. But such keyboards cannot cure the inevitable wandering in pitch when one goes up and down by ideal intervals only.

It is comparatively easy to hear in sustained chords that the thirds of equal temperament are "out of tune," because they are slightly higher than just tuning. However, experiments by Max Mathews show that some people like major triads with out-of-tune thirds; they sound "brighter." Good musicians can distinguish equal temperament from just tuning in listening to the successive notes of a scale. But the tuning of the piano is acceptable to most, and violinists in orchestras ordinarily seem to approximate just intervals no more closely than the piano does.

Sometimes just intervals are important, as in some passages in early choral music, and in barbershop quartets. Good singers will sing just intervals in such passages, and maintain the initial key through using unjust intervals where these are acceptable.

Tonal Centers of Scales

There is one other important characteristic of a scale. On what pitch does it start?

The diatonic C-major scale, represented by the white keys of the piano, consists of seven notes, or eight if we span a whole octave. Within the octave of the major scale, five of the intervals are whole tones (two semitones), and two (the third and the seventh) are semitones. The succession of intervals that we will encounter in going up the scale depends on what note we choose as the beginning of the scale. Sharps or flats (called chromatics) are added to preserve the sequence of whole tones and semitones, since this sequence defines the major scale.

Classical Greek music was based on a mathematically derived scale made up of tetrachords, groups of four descending pitches spanning a perfect fourth. The two middle pitches of each tetrachord could be chromatically altered; the unaltered version was referred to as diatonic. The entire collection of tetrachords, plus one added pitch at the bottom, formed a single two-octave scale. It probably always had as its center tone the A below middle C.

The (unrelated) medieval Church modes (sometimes erroneously referred to as the "Greek" modes) were based on ascending diatonic octaves. The tonic center of a mode is always the lowest pitch in the octave. The modes were codified in a process begun by Pope Gregory I in an attempt to standardize the great number of chants that had accumulated in early Christian liturgy (hence the name Gregorian chant). The Church modes flourished from about 800 to about 1500.

Although the Church modes largely fell into disuse in about the seventeenth century, two of them—which we now call the major and minor modes or scales—survived and became the tonal basis of Western European music. Some modern composers have written works or parts of works in a neomodal style. Various other types of chromatically altered scales are the basis of certain types of European folk music, most notably Eastern Europe and gypsy music.

If we start at C on the piano and play the white keys to C′, we play a major scale, in which the pattern of whole tones (1) and semitones ($1/2$) is 1, 1, $1/2$, 1, 1, 1, $1/2$.

If we start at A and play the white keys to A′, we play a minor scale, in which the interval pattern is 1, $1/2$, 1, 1, $1/2$, 1, 1. There are two other variants on the minor scale as well, in which the last three intervals in the pattern can be 1, 1, $1/2$, or even $1/2$, $1 1/2$, $1/2$ produced by using (in the key of A minor) F♯, or G♯, or both, instead of F and G.

We may observe that the major scale (think of the scale that starts on C) has major triads on the tonic, the fifth, and the fourth, all important tones of the scale. In the first minor scale mentioned (think of the white keys starting on A) the triads on the tonic, the fifth, and the fourth are all minor triads, a sort of complement of the case of the major scale.

In this book, the central concern is with the science of musical sound. This chapter has noted that the musical intervals produced by instruments can be tuned by listening for beats between the harmonics of two notes that are sounded simultaneously. We have seen that one cannot choose the frequencies of a seven-tone (or twelve-tone) scale in such a way that the ratios of the frequencies of the notes of the intervals are ratios of integers. We have seen that equal-tempered tuning, based on semitones of frequency ratio 1.059463, gives excellent approximations to ideal intervals, and that these approximations are the same for scales in all keys.

When we venture beyond the matters of intervals and tuning to discuss modes, we are wandering a little beyond the science of musical sound into the territory of music itself. This we cannot wholly avoid. But we will concern ourselves only with the two surviving modes: the major scale, because it is important to Rameau's ideas concerning harmony; and the minor scale, because it is hard to avoid.

5 Consonance and Dissonance

Melody is characteristic of almost all music. Strangely, so is the diatonic scale (or some subset of it), with intervals based on frequencies in the integer ratios 2/1, 3/2, 4/3, 5/4, 6/5, and so on. No doubt, musicians of many ages and cultures have plucked strings simultaneously and tuned their instruments by observing beats, as piano tuners do today. However, sounding notes together in what we conceive of as harmony is peculiar to Western music.

The notes of one instrument or several can be sounded together for a variety of purposes. In giving instructions for the representation of cannon shots in his piece *Les Caractères de la guerre*, written in 1724, François Dandrieu recommended that the harpsichordist "strike the lowest notes on the keyboard with the entire length of the hand." By 1800, another French composer recommended striking the lowest three octaves with the flats of both hands in his rendition of the sound of a cannon.

Whatever we think of as the essence of harmony, it does not lie in such sounds, or in the near–tone clusters of Percy Grainger, or in the tone clusters of Henry Cowell, or even in the haunting reminiscence of church bells that Charles Ives evokes in his *Concord* Sonata by instructing the pianist to strike simultaneously many black keys, softly, with a bar of wood (see Figure 5-1).

We think of harmony as notes that are sounded together smoothly and sweetly, or as rough-sounding or dissonant combinations of notes, full of tension, that miraculously resolve into a succeeding consonant chord. Beyond this, we think of well-known (and sometimes well-worn) progressions from chord to chord that serve as phrases or words of music. We

Figure 5-1 Page from score of Charles Ives's *Concord* Sonata, showing instructions for playing black keys with a stick.

think of modulation, the shifting from key to key, which is sometimes forthright, sometimes elusive or ambiguous.

About the harmonic language of music this book has little to say. Our concern will be chiefly with the consonance or dissonance one experiences when two, three, or more notes are sounded together. It is these experiences of consonance and dissonance that underlie the evolution of the musical theory of harmony.

Consonance, Dissonance, and the Critical Bandwidth

Of consonance, dissonance, and chords, there can be several views. One view might be that we are culturally conditioned to regard certain combinations of pitches as consonant, others as dissonant. However, composers of our century present us with an extraordinary range of combinations of notes, and with some rules for making such combinations. Can consonance and dissonance spring from nothing more than rules and customs? As will be seen, in a sense they can.

I believe, however, that to look at concepts of consonance and dissonance as arising from rules is to look at things the wrong way round. Rather, I am sure that the rules and customs are based on experiences of consonance and dissonance that are inherent in normal hearing. Of course, musical training and sophistication will color the subjective experience. The trained ear hears much in harmony that escapes the musically untrained. Sometimes the trained ear hears things that aren't there, for a musical friend of mine hears G♯ and A♭ as different on the piano. All this we will consider later.

Let us return to the piano tuner of Chapter 4. Beats and "roughness" are phenomena that are crucial to consonance and harmony. We are already acquainted with beats. We know that the piano tuner tunes intervals for the absence of beats, or else for the number of beats per second necessary for the equal-tempered scale. In the nineteenth century, Helmholtz tried to explain consonance and harmony entirely in terms of beats. He thought that intervals were consonant if there were no (or few) beats between their partials. To explain dissonant intervals, he proposed that partials of different tones were so close together in frequency that the beating between them was perceived as dissonance.

The work of Rainier Plomp and of others at the Institute for Perception Research (IPO), which Jan Schouten founded in the Netherlands, has shown that this is too simple a view. Slow beats do not give a sense of dissonance, but merely a rising and falling of amplitude. Further, as we gradually separate the frequencies of two sine waves or "pure tones," we hear a disagreeable roughness even when the frequencies are so far apart that we no longer distinguish beats. The range of frequencies in which we hear beats or roughness is called the *critical bandwidth* and is shown in Figure 5-2.

The critical bandwidth is an important experimental fact of hearing. To some degree, in listening to sounds we can tune in on a narrow band of frequencies, much as we tune in to a radio channel. When frequency components are separated by more than a critical bandwidth, we can hear them separately (this is called *hearing out*). But frequency components that lie within a critical bandwidth interact, and give us sensations of beats, roughness, or noise.

The critical bandwidth is important in the perception of loudness, in the perception that a sound is noise, and in the *masking* or hiding of one sound by another. In essence, the critical bandwidth results from the way that the ear resolves frequencies. At low and moderate sound levels, frequency components lying farther apart than a critical bandwidth send signals to the brain over separate nerve fibers, but frequency components lying within a critical bandwidth send a mixed signal over the same fibers.

The graph in Figure 5-3 shows the relation between critical bandwidth and consonance for two pure tones in another way. We can see from it that the maximum dissonance occurs at about a quarter of a critical bandwidth. With greater frequency separations, consonance also increases and becomes almost perfect for *all* separations greater than a critical bandwidth.

In order to know what frequency intervals between pure tones are consonant, we must know how the critical bandwidth varies with fre-

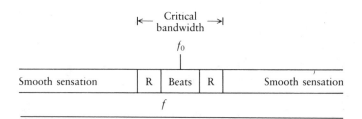

Figure 5-2 Pure tones (sine waves) that are close enough together in frequency give rise to audible beats. Even when the frequencies of such tones are too far apart for us to distinguish beats, we can still hear a certain "roughness." If the frequencies differ still more, we hear each tone separately and smoothly.
Imagine one pure tone to have a frequency f_0 and another to have an adjustable frequency f. As we vary f from much below f_0 to much above f_0 we pass from smooth to rough to beats to rough to smooth again, as shown. The range of frequencies within which we hear roughness or beats is called the *critical bandwidth*.

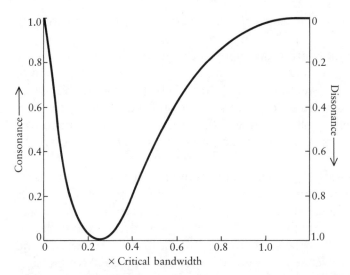

Figure 5-3 Plomp's curve representing the consonance of two pure tones sounded together as a function of their frequency separation (expressed as a fraction of their critical bandwidth). The tones are consonant if the frequency separation is so small that we hear slow beats, and consonant if the frequency separation is a critical bandwidth or greater, so that we hear the tones separately. The consonance and dissonance scales are chosen arbitrarily, to give consonance increasing from 0 to 1 (maximum) upward on the left, and, conversely, dissonance increasing from 0 to 1 downward on the right. The horizontal scale at the bottom represents frequency separation in fractions of a critical bandwidth. While this curve is derived from experimental data, its accuracy is not high.

quency. This is shown in Figure 5-4. Vertically it shows, for a critical bandwidth, the difference between the two frequencies that lie at its edges, f_1 and f_2. Horizontally the graph shows the average value of two frequencies, that is, $(f_1 + f_2)/2$.

We see that, for most of the frequency range shown, the critical bandwidth lies between a minor third and a whole tone. For frequencies below 440 Hz (the A above middle C), the critical bandwidth is larger. Thus we might expect that, in order to be consonant, notes of low frequencies that are sounded together would have to be more widely separated than notes of high frequencies that are sounded together. Indeed, in piano music the notes of the chords in the bass are commonly more widely separated than the notes of the chords in the treble. However, for many musical sounds lower frequencies cannot be very important to consonance, since we hear mostly the higher partials. This is partly because

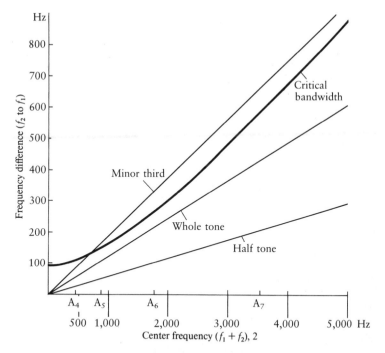

Figure 5-4 The width of a critical bandwidth as a function of frequency. Above a frequency of 500 Hz the critical bandwidth is roughly proportional to frequency. The frequency differences for three intervals are shown. A good approximation is that, if two pure tones are separated by a minor third or more, they will sound consonant together.

a good deal of the energy is in the higher partials (almost all when we listen to a transistor radio), and partly because the ear is more sensitive to higher frequencies than to lower frequencies.

A good rule of thumb is that pure tones less than a minor third apart are dissonant, but tones a minor third or more apart are consonant; so, for pure sine tones, *any* interval greater than a minor third will be judged as consonant, however odd the ratio of frequencies. But, of course, this isn't so for musical tones, such as piano tones, which have many harmonic partials. And musicians tend to judge the consonance or dissonance of pairs of sinusoidal tones by first recognizing the musical interval, and then calling the pair of tones consonant or dissonant on the basis of their past experience with pairs of nonsinusoidal tones and on the basis of what they have been taught.

Figure 5-5 shows the result of a calculation of the relative dissonance (or consonance) according to Plomp for two complex tones, each of which

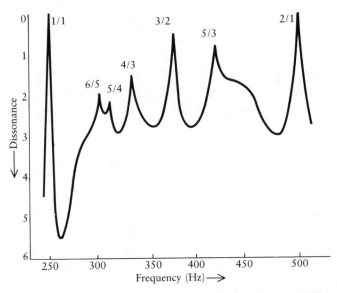

Figure 5-5 If we know the critical bandwidth, which depends on frequency, and use the curve in Figure 5-3, we can calculate the consonance of a pair of tones, each having six harmonic partials, as a function of the frequency separation of their fundamentals. The curve in this figure is based on such a calculation. The frequency of the fundamental of the lower tone is fixed at 250 Hz, and that for the upper tone varies from a little below 250 Hz to a little above 500 Hz (an octave above 250 Hz). The ratios above the peaks of consonance show the frequency ratios of the traditionally consonant intervals: 1/1, unison; 6/5, minor third, 5/4, major third; 4/3, fourth; 3/2, fifth; 5/3, sixth; 2/1, octave.

consists of a fundamental frequency and six harmonic partials. Here we see peaks of consonance at the familiar musical intervals. Whether or not the result represents exactly what we hear, it exhibits features characteristic of harmonic relations. The consonance of the octave is the easiest to explain.

If we designate the fundamental of the lower tone as f_0, the partials of the lower note of the octave are f_0, $2f_0$, $3f_0$, $4f_0$, $5f_0$, and $6f_0$, and those of the upper note are $2f_0$, $4f_0$, $6f_0$, $8f_0$, $10f_0$, and $12f_0$; so the partials either coincide or are well separated.

Next to the octave, the fifth is the most nearly consonant interval, traditionally and as shown in Figure 5-5. Figure 5-6 indicates why this is so. In this figure horizontal distance is a measure of frequency separation in octaves; that is, frequency components with ratios of 2 to 1 are one octave apart. The six harmonics of the lower tone are shown as short vertical lines above the axis, labeled 1 through 6. The six harmonic partials of the upper tone, a fifth or .58 octave above the lower tone, are shown as short vertical

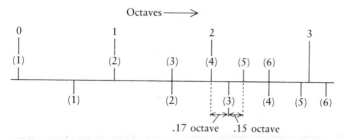

Figure 5-6 The proximity of the partials for the two tones in Figure 5-5, each with six harmonic partials, separated in frequency by the musical interval of a fifth. This interval is consonant because most of the partials of the two tones either coincide or are separated by more than a minor third. The third and sixth partials of the lower tone (shown above the horizontal line) coincide with the second and fourth partials of the upper tone (shown below the horizontal line).

lines below the horizontal line, labeled 1 through 6. Note that two partials of the upper and lower tones coincide. Most others are separated by more than a quarter of an octave. The third partial of the upper tone falls between the fifth partial of the lower tone (.15 of an octave away) and the fourth partial of the lower tone (.17 of an octave away). Thus the smallest separation of any two partials when two tones (each with six harmonic partials) a fifth apart are sounded together is about a whole tone (.17 octave). This is only a semitone less than a minor third (.25 octave).

We may note that the separation between the fifth and sixth partials of each tone is .26 octave, just greater than a critical bandwidth. If more partials are included in an individual tone, the higher partials will be less than a quarter of an octave apart, and the tone will have a sort of internal dissonance. We hear this as the buzzy quality of crude, electronically produced sounds, such as square waves and sawtooth waves. We also hear it in the jangly quality of the harpsichord. The distinction between consonant and dissonant intervals is somewhat muddied for pairs of internally dissonant tones.

We could make diagrams like that in Figure 5-6 for other intervals, such as the minor and major thirds, the fourth, and the sixth. In each case we would find some coinciding partials, and some partials closer (measured in fractions of an octave) than any in the diagram shown. But for dissonant intervals, such as a half tone, a whole tone, a tritone (an augmented fourth, or six semitones), or a seventh, many partials would lie much closer together than a quarter octave; sounded together, notes separated by these intervals sound rough and dissonant.

Perception of Consonance and Dissonance

The foregoing may seem to imply that we have found the root of consonance and harmony in the avoidance of partials that lie too close together in frequency. If partials are too close, they will beat or sound rough. For tones made up of many (say, six) harmonic partials, we can avoid serious beats or roughness by sounding together notes whose fundamental frequencies have integer ratios, such as 3/2 (fifth), 5/3 (sixth), 4/3 (fourth), 5/4 (major third), or 6/5 (minor third). Have we indeed found the unassailable basis of consonance? Perhaps.

For a pair of sine waves, naive listeners will judge *any* interval greater than a critical bandwidth as consonant; this is true even for the seventh, and even for intervals with irrational ratios of frequencies. However, trained musicians will dutifully identify thirds, fourths, fifths, and sixths, and report them as consonant, but will report sevenths as dissonant.

For complex tones made up of many partials, some musicians give strange judgments. I have heard one characterize an isolated deceptive cadence, shown at the right in Figure 5-7, as dissonant. Both final chords in the authentic cadence (left) and the deceptive cadence (right) are consonant, but the final chord in the deceptive cadence sounds unexpected and "wrong," as you can hear by playing it on the piano.

Much simple music closes with a cadence from the dominant seventh to the tonic (see Figure 5-8). The dominant seventh is an acoustically dissonant chord. Is this why the cadence is effective and identifiable? Perhaps historically. Producing sounds with computer, Max Mathews and I carefully deleted all partials closer than a quarter octave in both chords, rendering the dominant seventh chord consonant, and the tonic chord somewhat more consonant than is "natural." A trained musician identified the chords properly, even though the dominant seventh was no longer acoustically dissonant. The clue was most likely the half step from F to E, in addition to the half step from the leading tone (B) to C.

Figure 5-7 An authentic cadence (left) and a deceptive cadence (right), which ends on a chord of the sixth (submediant) instead of on a tonic chord. The final chord of the deceptive cadence is consonant, but the progression sounds odd and not final.

Figure 5-8   A cadence that goes from the dominant seventh, an acoustically dissonant chord, to the tonic, a consonant chord.

Figure 5-9   This score is the beginning of *Stabat Mater* by Palestrina, cited by Helmholtz. The passage is consonant, but, to Helmholtz, the succession of "chords" seemed odd because they did not clearly establish a key.

half step from F to E, in addition to the half step from the leading tone (B) to C.

Nonetheless, we will see in the next chapter that beats, or lack of beat, does play a crucial role in musical perception. If either the diatonic scale or the harmonic partials essential to it is slightly tampered with, overall musical or consonant effect is destroyed.

How we "hear" musical sounds depends partly on cultural bias and musical training. But it is clear there is a basis for consonance and dissonance in the very nature of human hearing, since the critical bandwidth is related to the way that the auditory nerves transmit messages to the brain. Nevertheless, harmony consists of more than an avoidance of dissonance. We can see this from Helmholtz's remarks about the beginning of Palestrina's eight-part *Stabat Mater* (Figure 5-9). Of this passage, which is certainly consonant, Helmholtz says:

> Here, at the commencement of a piece, just where we should require a steady characterization of the key, we find a series of chords in the most varied keys, from A major to F major, apparently thrown together at haphazard, contrary to all rules of modulation. What person ignorant of ecclesiastical modes could guess the tonic of the piece from this commencement? As such we find D at the end of the first strophe, and the sharpening of C to C♯ in the first chord also points to D. The principal melody, too, which is given in the tenor, shows from the commencement that D is the tonic. But we do not get a minor chord of D till the eighth bar, whereas a modern composer would have been forced to introduce it in the first good place in the first bar.

Helmholtz then observes how different the system of Church modes was from our system of major and minor keys. Someone who is unfamiliar with sixteenth-century polyphony might try to interpret this music as D major or D minor or in some other key.

We will explore the ideas of consonance and harmony further in Chapter 6.

6 Consonance, Dissonance, and Harmony

*F*or many years I was convinced that the work of Helmholtz, as augmented by Plomp, had established the physical and psychophysical basis of both consonance and harmony. Things seemed so simple. Sounds were consonant when no (or few) frequency components (partials) fell within the same critical bandwidth. This concept explains both the traditionally consonant intervals, such as the octave and the fifth, and the traditionally dissonant intervals, such as the second or the seventh.

In conventional counterpoint of any number of voices, a dissonance occurs only when any one voice "offends" against any other. Likewise, in chords consonant by conventional standards of harmony, all the intervals among notes meet Helmholtz's criteria to some degree. Thus I considered harmony an outgrowth of counterpoint.

A Synthetic Scale

Influenced by the ideas of Helmholtz and Plomp, in 1966 I proposed to expand the effects of musical harmony by synthesizing an entirely new scale. This scale consisted of eight equidistant notes spanning the octave, as shown in line B of Figure 6-1. The frequency ratio between any two successive notes in this scale is 1.0905; so the frequency ratio between the first and the third notes of the scale is 1.1892, which is the ratio for an equal-tempered minor third.

In effect, I had used C, E♭, G♭, and A as notes of my scale, and then had added a new note 1.5 semitones above each of these notes.

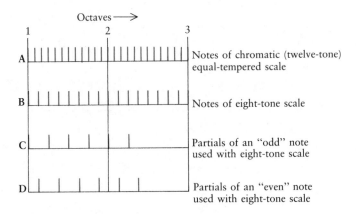

Figure 6-1 Frequencies in this illustration are measured in octaves, not in Hertz. Thus, the frequency of octave 2 is twice the frequency of octave 1, and the frequency of octave 3 is four times that of octave 1. The vertical bars along horizontal line A show the fundamental frequencies (first partials) of the notes of the equal-tempered chromatic (twelve-tone) scale. The vertical bars along line B show the frequencies of the first partials of the eight-tone scale that I used in the *Eight-Tone Canon*. The vertical bars along line C show the frequencies of the first six partials of an "odd" note, the first note in the eight-tone scale. The vertical bars along line D show the first six partials of an "even" note in the eight-tone scale. All partials of an even *or* of an odd pitch are one-fourth of an octave apart. Partials of odd pitch coincide; partials of even pitch coincide. But the partials of an even note and an odd note are only an eighth of an octave apart, a very dissonant interval.

Not only did I devise a new scale, but for it I synthesized new tones in which all partials except the octave partials were nonharmonic. In fact, the partials of each tone were simply the frequencies of every other higher note in the scale; so all partials were an (equal-tempered) minor third apart, as shown in lines C and D of Figure 6-1. To a musician, the tones and their partials sounded like diminished seventh chords. To me, they sounded rather fruity but not dissonant, in accordance with the theories considered in Chapter 5.

Which of these new tones are consonant when sounded together? Suppose that we distinguish the notes of my scale as *odd* and *even*; that is, the first, third, fifth, and seventh notes are odd, and the second, fourth, sixth, and eighth notes are even. The first six partials of an odd note are shown in line C of Figure 6-1; the first six partials of an even note are shown in line D. Successive odd notes are a quarter of an octave (a minor third) apart; so are successive even notes. This is the same interval as that between the partials in the tones that I synthesized. But adjacent odd and even notes in the scale are an eighth of an octave apart. An examination of the figure above and a little reflection will show that the laws of consonance for this scale and the tones used with it are very simple. When any two odd notes are sounded together, all partials of the two tones that are sounded coincide with one another, or miss completely — the higher note has no partial that coincides with the lowest partial of the lower note. The same is true for two even notes. Hence any pair of odd notes, or any pair of even notes, will be just as consonant as the tones themselves. However, if we sound an odd note and an even note together, those two notes and some of their partials will lie an eighth of an octave apart. An eighth of an octave is less than a critical bandwidth; so odd notes sounded with even notes should sound very dissonant. Indeed, I found that they did.

Thus, for this eight-tone scale, the laws of consonance and, I thought, of harmony were extremely simple. Any and all odd notes go together; any and all even notes go together; odd and even notes sounded together result in dissonance.

To illustrate all this, I composed a four-voice *Eight-Tone Canon*, which was included in a Decca record, *The Voice of the Computer* (DL 71080), now out of print. In this canon I attempted to use transitions from dissonance to consonance in a "musically effective" way. I thought that the piece sounded pretty good, but in writing contrapuntally I had really evaded a clear test of harmony as we commonly understand it.

Stretched Octaves and Partials

In 1979 I had an opportunity to spend a month in Paris at Pierre Boulez's IRCAM (Institute for Research and Coordination of Acoustics and Music). Emboldened by the ideas of Helmholtz and my *Eight-Tone Canon*, I resolved to attempt what seemed a crucial experiment, one that Frank H. Slaymaker had performed in part in 1970. This experiment involved the playing of conventional chords and music with "stretched" octaves, or pseudo-octaves, synthesized electronically. This concept of stretching is illustrated in Figure 6-2.

Suppose that, instead of the normal octave, with a frequency ratio of 2, we use a pseudo-octave whose ratio of frequencies is 2.4. Suppose that we also stretch, in a consistent way, the intervals between *all* frequency components of *all* partials of *all* notes that are sounded. A single tone in this system will no longer have harmonic partials (as in part A of the figure). Rather, it will have nonharmonic partials, whose frequencies are not integer multiples of the frequency of the first partial (as in part B). The successive (equal-tempered) semitones of the stretched chromatic scale will

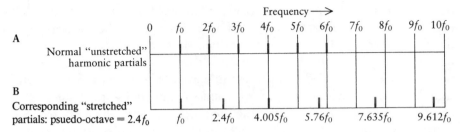

Figure 6-2 Frequency in hertz is measured from left to right. The vertical bars along the upper horizontal line (**A**) show the frequencies of the harmonic partials of a "normal" tone whose fundamental is f_0. The vertical bars along the lower horizontal line (**B**) show the frequencies of the nonharmonic partials of a uniformly "stretched" tone in which the pseudo-octave has a frequency of 2.4 f_0, rather than the frequency 2 f_0 of a true octave. Note that the spacing between successive harmonic partials is always the same, but spacing between successive stretched bars increases with increasing frequency of the partials.

have frequency ratios of 1.0757, instead of 1.0594 as in the usual chromatic scale.

The result of this stretching of both the scale and the partials of the tones used with the scale is very simple. Suppose that we play the same music twice, first as *normal* (that is, unstretched and with harmonic partials) and then ~~with~~ stretched and with stretched partials. If any two partials of two notes coincide in frequency in the unstretched version of the music, they will coincide also in the stretched version. Partials that don't coincide in frequency in the normal music will be a little farther apart, and hence a little more consonant, in the stretched version than in the normal version. Therefore, according to Helmholtz and Plomp's theories, any combination of notes that is consonant in the normal music will be consonant, or a little more consonant, in the stretched music.

What did the stretched music sound like? We played a stretched version of *Old Hundredth* in four-part, note-against-note harmony for Pierre Boulez. He said that the only structure he could discern was the two fermatas (somewhat longer notes) at the middle and end of the piece. All familiar harmonic effects had vanished, and he apparently discerned no melody.

Those who heard the unstretched version first (we used both *Old Hundredth* and the *Coventry Carol*) could recognize the melody, and perhaps distinguish the lowest voice, whose first partial was the lowest frequency present. However, it was hard, if not impossible, to follow the inner voices.

One of the most striking harmonic effects is the "finality" or "closing" quality of the *cadence*, a progression from the dominant chord (such as B–D–G in C major) to the tonic chord (C–E–G), in which we go from the leading tone (the seventh, or B) to the tonic, C. Such a progression sounds like "The End" to anyone familiar with Western music. In the stretched harmony, musicians could not tell a cadence from an "antica-dence" that went from the tonic to the dominant! Furthermore, they could not tell an authentic cadence from a deceptive cadence that ended on the chord of the sixth (C–E–A) rather than on the tonic. Neither seemed more final; both seemed equally strange and equally consonant (or dissonant).

Some experiments carried out at about the same time at Stanford by Elizabeth Cohen as a part of her doctoral work shed light on this. Among other things, she asked people to what degree a stretched tone seemed to be fused with its partials. That is, did they hear a stretched tone as a single sound, or simply as a collection of different frequency components? She found that, when the octave was stretched more than about 5 percent, the partials were not heard as a single tone of a particular timbre. Rather, listeners heard the collection of stretched partials as distinguishable tones of different frequencies. The higher-frequency partials were not individu-ally distinct, but stood apart from the lower frequencies as a group.

This was something of a surprise to me, for many sounds with non-harmonic partials, such as the sounds of bells, gongs, drums, or knocking on wood, are heard as distinct, identifiable timbres. In part, such natural sounds are "held together" by a common variation of different frequency components over time, such as a sharp attack and concurrent decay in amplitude, or, in the case of musical sounds, common small variations in the amplitudes or pitches of different frequency components (a common vibrato). The ear does not analyze these sounds into separate components; yet that is just what our hearing did to the computer-produced tones with considerably stretched partials. The ear simply refused to hear such a collection of independently produced, unvarying partials as a distinct tone with a single pitch and a single distinctive quality. Instead, the ear picked these stretched tones to pieces.

This, I think, accounted for the confused impression made by four-part harmony with stretched chords. Because the ear could not identify the tones as single sounds, several tones sounded together were heard, not as a chord, either consonant or dissonant, but rather as a sort of mush of sound. That is why Boulez could hear no structure in a stretched *Old Hundredth*, even though all the mathematical structure present in the normal version was present in the stretched version, and, by the rules of Helmholtz and Plomp, consonance and dissonance were preserved.

In 1987, IPO issued a wonderful disc by Houtsma, Rossing, and Wagenaars entitled *Auditory Demonstrations*. Tracks 58 through 61 illus-trate the effects of a moderate stretching (octave of 2.1 rather than 2.0) of scale frequencies and/or partial spacings. Part of a Bach chorale is played with synthesized tones. When neither scale nor partial frequencies are stretched, we hear the intended harmonic effects. When the scale is unstretched but the partial frequencies are stretched, the music sounds awful. Clearly, melodic intervals in the ratios of small whole numbers are in themselves insufficient to give Western harmonic effects. Further, when the scale is stretched but the partial frequencies are unstretched, again the music sounds awful. The harmonic effect does not lie simply in tones with harmonic partials. But when *both* the scale intervals and the partial frequencies are stretched a little, the music sounds slightly odd but good.

As we have already noted, when both scale intervals and partial spacings are stretched by the same small amount, partials associated with different pitches that fell on or near one another for unstretched intervals will fall on or near one another for stretched intervals and partials. If the stretching is small enough, something of the harmony is preserved.

This demonstration appears to show that the coincidence of near-coincidence of partials we find for normal (harmonic partials) musical sounds and for consonant intervals (with frequency ratios in the ratio of small integers) is a necessary condition for Western harmonic effects — in Bach at least. When this coincidence or near-coincidence is preserved, harmonic effect is preserved, even with small deviations from the ratios of small integers in scale intervals and in frequencies of partials.

Whatever the case may be, for Western ears there appears to be something musically magical about harmonic partials and the diatonic scale.

Virtual Pitch

When I visited the Philips laboratory in Eindhoven, Holland, shortly after World War II, Jan Schouten, an ingenious physicist, showed me a fascinating experiment. He had constructed a sort of optical siren (Figure 6-4) by means of which he could produce sounds with various waveforms. Using this, he produced sounds with harmonically related partials of frequencies $f_0, 2f_0, 3f_0, 4f_0$, and so forth. Then, by proper adjustments, he could cancel out the fundamental frequency f_0. I could hear this fundamental frequency come and go, but the pitch of the sound did not change at all. In some way, my ear inferred the proper pitch from the harmonics, each separated from the next by the frequency f_0 of the fundamental.

Schouten's optical siren made use of a rotating disc with radial, narrow, equally spaced transparent slits. When these swept past a single stationary radial transparent slit they let through periodic pulses of light. These were translated into electric pulses by a photoelectric cell. The

Figure 6-3 Jan Schouten.

rotating transparent slits also swept past a transparent region shaped so as to produce a continuous sine wave from a second photoelectric cell as the radial transparent slits swept by. The phase of this sine wave could be shifted by rotating the sine-wave-producing region about the axis of rotation. The amplitude of the sine wave could be controlled by a gain control. Thus, the phase and amplitude of the sine wave could be adjusted so as to cancel the fundamental frequency of the sequence of pulses produced as the radial transparent slits swept past the stationary radial transparent slit.

I later found that Schouten had published an account of this work in 1938, and in a paper published in 1940 he gave the name *residue pitch* to the pitch correctly heard in the absence of the fundamental frequency. This pitch has since been called *periodicity pitch* or *virtual pitch*, a term introduced by Ernst Terhardt.

Schouten's observation should not surprise anyone who has listened to a pocket transistor radio. The speaker of such a radio is so small that the fundamental frequencies of all the lower tones are too weak to be audible; yet we hear the proper pitches, however tinny the music may sound. Perhaps the best name for what we hear is simply *musical pitch*.

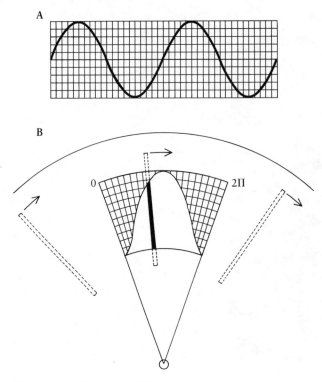

Figure 6-4 Diagram of Schouten's wave siren.

Had Helmholtz not created, or perpetuated, a myth that the pitch of a musical sound is conveyed by the presence of the fundamental frequency, Schouten's result would have seemed less surprising. For musical tones of very high pitch, such as those of the glockenspiel, the higher-frequency partials or overtones don't contribute much to loudness; the chief musical impression, timbre as well as pitch, is derived from the fundamental. For tones toward the low end of the piano keyboard, the case is quite different. The omission of the component of fundamental frequency has an inappreciable effect on the pitch *or* the timbre. The same is true for the second partial, the second harmonic. The third harmonic has some effect on timbre. Clearly, toward the low end of the piano keyboard musical pitch is conveyed by salient harmonics or partials, not by the fundamental.

This is consistent with the procedure of tuning pianos and organs. In the tuning process, the twelve tones of a central octave are tuned by an adjustment of beats between tones of consonant intervals. Then the upper and lower registers are tuned by octaves, that is, by beats between harmonics of the lower tone and the fundamental and harmonics of a tone an octave higher. In the higher tone there is no frequency component for the fundamental component of the tone an octave lower to beat with.

At the high end of the range of musical pitch as defined by the piano keyboard, the lowest or fundamental frequency component is all-important to musical pitch. At the low end of the keyboard, the fundamental frequency component is unimportant to pitch, and to timbre as well. At the middle of the keyboard (around middle C) the omission of the fundamental changes the timbre (making the tone brighter) but does not change the pitch. In the mid-keyboard range, either the lowest partial, the fundamental, or the higher partials can convey pitch.

Psychoacousticians have studied the residue or virtual pitch, or musical pitch for a wide variety of combinations of sinusoids, and some of the results are curious.

When a few partials have equal spacings but are not exactly harmonics (integer multiples) of any lower frequency, the virtual pitch is some frequency of which the partials are *approximately* harmonics. Thus, the frequencies 820, 1020, and 1220 Hz can be heard as the fourth, fifth, and sixth harmonics of 204 Hz (816, 1020, and 1224 Hz); a pitch of about 204 Hz will be heard when these frequencies are sounded together.

Orderly collections of sinusoids do not always produce a virtual pitch equal to that of the missing fundamental. Sinusoids that are exclusively odd-numbered harmonics of a missing fundamental do not produce the pitch of that fundamental. But the seventh, ninth, and eleventh harmonics of 16 Hz, that is, 112, 144, and 176 Hz, are approximately the fourth, fifth, and sixth harmonics of 28.8 Hz (115.2, 144, and 172.8 Hz). They produce a pitch of about 28.8 Hz when sounded together.

Successive harmonics may or may not be heard as a virtual pitch of the missing fundamental. In generating combinations of sine waves with a Yamaha DX7, I observed that toward the low end of the piano keyboard, three sinusoids with the frequencies of C, E, and G sound like a single tone with a pitch two octaves below C, of which they are the fourth, fifth, and sixth harmonics. But toward the center of the keyboard, C, E, and G sounded together sound like a major triad, with C the lowest tone. However, still near the center of the keyboard, if we add G to C and C′, the pitch drops an octave below C, a pitch of which C, G, and C′ are the second, third, and fourth harmonics.

Nonetheless, the ear has a strong tendency to ascribe a single pitch to a collection of tones whose frequencies are integer multiples of a common frequency, even though that frequency itself and some of its integer multiples are absent. Furthermore, we can have a sense of pitch and unity of sound even when the frequency intervals between successive higher partials are not exactly equal. This is how we ascribe a pitch to bells. This pitch is not the frequency of the lowest partial (the *hum tone*), but rather an average of the frequency separations between some higher partials of the sound of the bell.

Rameau and the Fundamental Bass

In his *Traité de l'harmonie* (*Treatise on Harmony*), first published in 1722, Jean-Philippe Rameau attributed the distinct character of a major triad to a fundamental bass (*basse fondamentale*). What is this fundamental bass?

A major triad is a chord made up of a note of fundamental frequency f_0; the major third above it, which has a fundamental frequency $(5/4) f_0$; and the fifth above it, which has a frequency $(3/2) f_0$; for example, C–E–G. As Figure 6-5 shows, the harmonics of C have frequencies

$$f_0, 2f_0, 3f_0, 4f_0, 5f_0, 6f_0$$

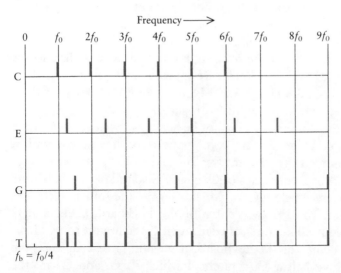

Figure 6-5 In this figure, the vertical bars along the horizontal line C indicate the frequencies of the first six harmonic partials of C; those along line E show the frequencies of the first six partials of E; and those along line G show the first six partials of G. Together, C, E, and G form a major triad. The frequencies of partials of all three pitches are shown together along line T. All the partials of all the pitches of the triad are integer multiples of a frequency $f_b = f_0/4$.

The harmonics of E have frequencies

$$(5/4)f_0, \ 2(5/4)f_0, \ 3(5/4)f_0, \ 4(5/4)f_0, \ 5(5/4)f_0, \ 6(5/4)f_0$$

The harmonics of G have frequencies

$$(3/2)f_0, \ 2(3/2)f_0, \ 3(3/2)f_0, \ 4(3/2)f_0, \ 5(3/2)f_0, \ 6(3/2)f_0$$

We see that all the partials of the notes present in the chord are integer multiples of the frequency

$$f_b = (1/4)f_0.$$

We can therefore write the harmonics of C as

$$4f_b, \ 8f_b, \ 12f_b, \ 16f_b, \ 20f_b, \ 24f_b,$$

the harmonics of E as

$$5f_b, \ 10f_b, \ 15f_b, \ 20f_b, \ 25f_b, \ 30f_b,$$

and the harmonics of G as

$$6f_b, \ 12f_b, \ 18f_b, \ 24f_b, \ 30f_b, \ 36f_b.$$

In the bottom row of the figure, the vertical bars indicate the frequencies of the first six partials of C, E, and G. Some of these coincide. Furthermore, many successive partials of f_b are present in this triad. The frequencies $4f_b$, $5f_b$, and $6f_b$ are successive harmonics of f_b. So are $15f_b$ (the third harmonic of E) and $16f_b$ (the fourth harmonic of C), as well as $24f_b$ (the sixth harmonic of C) and $25 f_b$ (the fifth harmonic of E). In listening to this triad, should we not therefore hear a pitch two octaves below C, the root of the chord, corresponding to $f_b = f_0/4$?

Rameau could have heard this fundamental bass, two octaves below the root of the chord, when he listened to major chords. However, he regarded notes an octave apart as essentially identical. He therefore assumed as the fundamental bass the actual lowest note, or the root, of the chord (in our case, C) when it is arranged in thirds — rather than the C two octaves below.

Rameau's next, inevitable conclusion was that what we now call various inversions of a chord are the same chord because they have the same fundamental bass. That is, E–G–C and G–C–E are the first and second inversions of C–E–G, and we regard them, as Rameau did, as essentially the same chord. Before Rameau, they were named differently and regarded as distinct chords. Rameau thus drastically reduced the number of "different" chords that a musician had to learn.

If the distinct character of major chords lies in their having a fundamental bass, so that the partials of all notes present in the chord are integer multiples of this fundamental-bass frequency, then we can easily understand the lack of harmonic effect in sufficiently stretched chords. When we stretch a chord, the partials of the stretched tones of which it is composed are not integer multiples of any one frequency. If we stretch the scale and the tones enough, we destroy the fundamental bass. The ear hears nothing to characterize the collection of stretched notes. Indeed, as we have seen, there is nothing that enables the ear to attribute a distinct pitch to the stretched tones with which we try to make up a stretched musical chord.

Although greatly stretched tones seem to be vague collections of sinusoidal components rather than fused sounds with distinct pitches, if we play a stretched scale of stretched tones, we do get a sense of rising pitch. Melody is more rugged than harmony. Melodies played in a stretched scale with stretched partials are easily recognizable. It is only harmony that is confused or lost altogether when the stretching is extreme.

As we have seen, Helmholtz correctly explained the basis of acoustical consonance. Plomp elaborated on his work, discovering that when too many partials lie within a critical bandwidth, a sound will be rough or dissonant, whatever other qualities it may have. This concept adequately explains many things, including the fact that, although melodies played on a carillon sound acceptable, harmony on a carillon sounds discordant. When "consonant" chords are played on a traditional carillon, many nonharmonic partials lie close together in frequency.

In Chapter 5 we noted that single notes as well as chords can have a dissonant character: the jangly sound of a note played on a harpsichord, or the buzzy quality of an electronically generated sawtooth wave or square wave. But there is more to harmony than this sort of consonance and dissonance.

Indeed, there is more to harmony than major triads. Rameau himself had trouble explaining minor triads, which cannot be derived from partials. Before about 1500 the final chord nearly always consisted of the root,

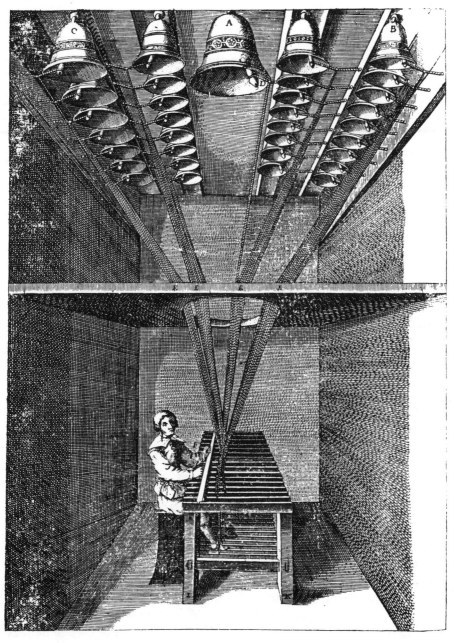

Figure 6-6 Carillon of the church of Notre-Dame at Anvers (Antwerp).

fifth, and octave—without a third, whether major or minor, as they were considered dissonant. Thirds gradually began to be regarded as consonant in the fourteenth century, but not consonant enough to appear in the final chord. Around 1500 the third began to be permitted in the final chord. The major third was preferred because it was regarded as more consonant than the minor third, and so was used in the final chord even when the rest of the piece was in the minor key. In this context, it was known as the "Picardy third."

The major triad and its fundamental bass, which was first recognized by Rameau, is a sort of scaffolding on which Western European composers built very elaborate structures of harmony. The major triads on the tonic, the dominant, and the subdominant (C, G, and F, for example) contain all the notes of the scale, as discussed in Chapter 4. Sounded in succession, these three chords indicate the key unambiguously.

As the sole resources of harmony, these three chords would be dull. If we add another third, we can get the dominant seventh (G–B–D–F), a most useful chord. The notes of a diminished seventh (B–D–F–A♭) are the same in four different keys; so this chord, which has an ambiguous and unstable character, is useful for modulating from one of these keys to another. A few extra notes will make any major or minor chord sound lusher or jazzier, or may lend it an intriguing touch of some other chord or some other key. Likewise, a chord can be rendered provocatively ambiguous by the omission of some of its notes.

We have seen that although conventional harmonic effect is preserved in tones *and* scales stretched a little, with greater stretching the tones themselves fall to pieces, and harmonic effect is lost. It appears that approximately harmonic partials, and scales embodying frequency ratios that approximate the ratios of small integers, are essential to harmony. Does this rule out scales other than the diatonic scale?

There has been a good deal of experimenting with scales. In his fine piece *Stria* (1977), John Chowning used partial spacings and pseudo-octaves in the ratio of the Golden Mean (approximately .618).

Later, I proposed a scale for use with tones having odd harmonic partials only. A *tritave* with a 3:1 frequency ratio (the ratio of the frequency to the lowest overtone, the third harmonic, to the frequency of the fundamental) is used instead of the octave. The analog of a major tetrad has the frequency ratios 3:5:7:9. A nine-tone scale containing six "major" triads and six "minor" triads can be chosen from a thirteen-tone "chromatic" scale with equal intervals in the ratio of $\sqrt[13]{3}$. The triads do sound chordlike. A few composers have produced attractive pieces using this scale, and singers have learned to sing and transpose in it. It appears to "work" with tones having successive harmonic partials as well as with tones having odd harmonic partials only.

Digital synthesis invites further experimentation with this and with other scales. Yet it remains possible that musical sounds with harmonic or nearly harmonic partials—that is, musical sounds that are periodic or nearly periodic—together with the diatonic scale with its diatonic harmony, are best adapted to human perception. We may note that both the spectrum of a periodic sound with successive harmonic partials and the diatonic scale involve the ratios of small integers, such as $2:1$, $3:2$, $4:3$, $5:4$, and $6:5$.

7 *Ears to Hear With*

*I*magine standing outdoors on a quiet evening with your eyes closed. What is that faint rustling sound down and to the right? Some small creature must be causing it. The susurrous sound above is the wind whispering among the dry leaves. In the distance we sense the approach of a rickety motorbike along the roughish road. It passes us with a rattle of decrepitude, departing on a slightly falling note. All of this, and more, we sense through two smooth, pencil-sized holes in our heads. And the wonderful sounds of music also reach our brains through the mechanism of the ears and the neural pathways from ear to brain.

In vision, two perspective views of the external world are imaged on the retinas. In hearing there is no such direct physical representation of external space. Yet, somehow, sound can give us a vivid impression of the world outside, of the saw biting through the wood, of the violin bow wakening the string—with a scratching noise if the player is a novice.

Let us begin our explication of what is known about hearing with ears.

Figure 7-1 shows the outer, middle, and inner ear. The visible outer ear, on the side of the head, is called the *pinna* (and the two ears are the *pinnae*). For many years, students of hearing thought that the pinna wasn't very important. In 1967 an independent scientist named Wayne Batteau showed that it is.

If you fold your ears over, or fill the convolutions with wax or modeling clay, and with your eyes closed listen to a nearby, high-pitched sound, such as the jingling of keys on a ring, you will find that you cannot judge the *height* of the sound, that is, tell whether such a sound is in front

102

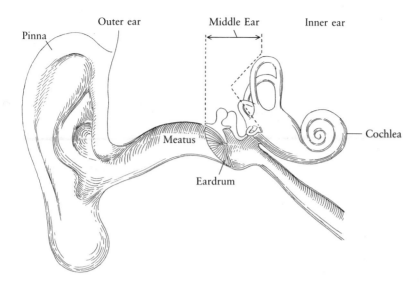

Figure 7-1 The outer ear, the middle ear, and the inner ear. The outer and middle ears are air filled; the inner ear is filled with various fluids. The loops at the upper left of the inner ear are the semicircular canals, which have to do with balance, not hearing. The cochlea is the snail-like organ at the right of the inner ear.

of you, above at an angle, or directly overhead. Because of our pinnae, the sensitivity of the ear to very high-frequency sounds changes markedly with both the direction of the sound source and the frequency of the sound. Somehow, this enables us to judge the height of the sound source.

The pinna channels sound waves to the auditory canal, or *meatus*, which is about 2.7 centimeters (1 inch) long. At the inner end is the ear drum, or *tympanic membrane*. The auditory canal acts as a rather broadband resonator, with a resonant frequency of about 2,700 Hz. Together with characteristics of the middle and inner ear, this broad resonance helps determine the frequency at which our hearing is most acute, which is about 3,400 Hz.

The eardrum divides the outer ear (the pinna and auditory canal) from the middle ear, which consists of three tiny bones. As the eardrum vibrates in response to sound waves, the bones convey these vibrations to a third portion of the ear, the inner ear. Figure 7-2 shows the bones of the middle ear. Their names are the hammer (*malleus*, which looks more like a club), the anvil (*incus*, which looks rather like a tooth), and the stirrup (*stapes*, which really does look like a stirrup). These bones, which are flexibly connected together, convey the eardrum's vibration to a membrane cover-

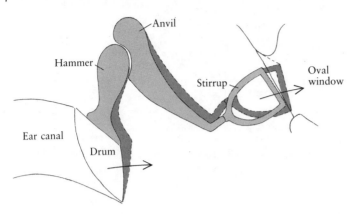

Figure 7-2   The three tiny bones of the middle ear convey vibrations from the eardrum to a membrane covering the oval window of the inner ear.

ing an opening called the *oval window*, which separates the air-filled middle ear from the fluid-filled inner ear.

In the inner ear the vibrations of sounds are sorted into overlapping ranges of frequencies. Then vibrations corresponding to different ranges of frequencies are converted into electric impulses that travel along different nerve fibers to the brain by means of a complicated, interlinked system of nerve pathways (see Figure 7-3). A good deal is known about the anatomy of these neural pathways to the brain and the "way stations" at which they are interconnected. Through animal experimentation, something is known of their function.

Because the vibrations of sound are sorted out into different, overlapping ranges of frequency, a sine wave, or a sinusoidal component of a complicated sound, will send nerve impulses toward the brain over a particular nerve pathway. Along the way, the response to one sound may interact with responses to others. But a distinct *tonotopic* mapping (*ton* for tone, *topic* for place or locality) of excitation of nerves, in response to the frequency of the sound heard, is preserved all the way up to the auditory cortex of the brain.

At each way station a band of firing neurons depicts a spectral analysis of the sound that caused the firing. If an animal hears a sequence of sine waves of increasing frequencies, the maximum responses at the way stations will shift in position in an orderly way, and across the surface of the auditory cortex also.

As we travel up the nerve pathways and to the auditory cortex, the tonotopic response will follow changes in the sounds heard more and more sluggishly, until at the cortex itself the response changes roughly at the rate

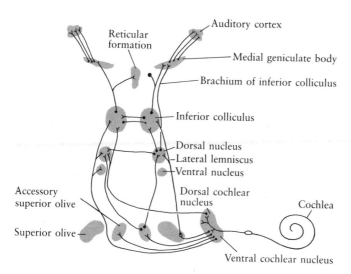

Figure 7-3 Neural pathways from the hair cells in the cochlea to the cerebral cortex, a part of the surface of the brain. Pathways from both left and right ears are interconnected at various "way stations." A good deal is known about the pathways and their interconnections, but less about the functioning of this complex system.

of the successive sounds of speech or the successive tones of music. Presumably, we make various *uses* of the spectral features, the frequency patterns of sounds, and these uses, quick or tardy, are made at different levels in the auditory system.

At the very lowest level, where the nerves from the two ears first come together, the response can follow several thousand changes a second. It is here that the comparison is made that enables us to sense the direction from which the sound comes by comparing time of arrival at the two ears. At the auditory cortex the slow response seems suitable to disentangling the successive sounds of spoken language.

Way stations must provide many complicated responses to complex, changing sounds, but our knowledge of these is limited.

The inner ear is important to our sense of balance as well as our hearing. Its bony case contains three semicircular canals, which do not take a direct part in hearing, but enable us to sense the attitude of the head, and hence of the body. Here we are interested in the auditory part of the inner ear.

The auditory part of the inner ear that is important to hearing has two functions: spectral analysis, or the sorting out of different ranges of frequency, and the excitation of nerves. Spectral analysis is carried out in the *cochlea*, a spiral, tapered tube like the inside of a snail shell. The function of the cochlea illustrates things that we have already learned in connection with resonance, filters, and spectral analysis.

The cochlea is shown on the right in Figure 7-1. The spiral tube of the cochlea makes two and a half turns and is about 3 cm (1.2 inches) long. It is about 0.9 cm (0.4 inch) in diameter at the beginning (basal) end, and about 0.3 cm (0.2 inch) in diameter at the far or apical end.

In order to be more easily understood, the cochlea is customarily drawn as a straight, tapered tube, that is, unrolled (as in Figure 7-4). The schematic cross section of the cochlea is shown at the right in the figure. W indicates the width of the basilar membrane. This springy membrane is shown arched downward, as it would be if the pressure in the fluid above was greater than the pressure in the fluid below.

The width, W, and the stiffness of the basilar membrane vary from the left (basal) end to the right (apical) end. The membrane is widest and laxest at the right end, narrowest and stiffest at the left end.

The longitudinal, or end-to-end, section of the unrolled cochlea is shown at the left in the figure. The basilar membrane does not extend all the way to the closed, apical end (on the right). Fluid above the basilar membrane can flow into the space under the membrane through an opening called the *helicotrema*.

At the basal end (at the left) are two windows in the bone that surrounds the tube of the cochlea. These are covered with thin, flexible membranes. The upper opening is called the *oval window*. The stirrup (see Figure 7-2) is connected to the membrane covering the oval window. When a sound wave causes the stirrup to move in, it pushes the membrane in and causes the fluid in the upper part of the cochlea to move to the right; when the sound wave causes the stirrup to move out, it pulls the

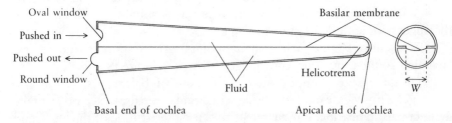

Figure 7-4 The cochlea "unrolled" and shown as a straight tapered tube. The cross section (at the right) is simplified to a circle with rigid inner projections that support the springy basilar membrane, whose width increases along the length of the cochlea, from the basal to the apical end. The longitudinal (end to end) section of the unrolled cochlea is shown at the left.

membrane out and causes the fluid in the upper part of the cochlea to move left.

The membrane that covers the round window below the basilar membrane isn't connected to anything. This membrane flexes in and out in accord with the pressure of the fluid under the basilar membrane.

If we pushed the oval window in slowly, fluid above the basilar membrane would flow to the right, pass through the helicotrema, flow to the left under the basilar membrane, and push the membrane over the round window out. Figure 7-4 shows the membrane over the oval window pushed in (by the stirrup) and the membrane over the round window pushed out by the motion of the fluid through the helicotrema. Actually, a sound wave causes the stirrup to move rapidly right and left, thus increasing and decreasing the pressure in the fluid above the basilar membrane. This causes a wave to travel along the membrane, through the cochlea from left to right. We can best visualize this wave as a traveling, up-and-down motion of the basilar membrane.

Let us assume that the stirrup moves left and right sinusoidally with a particular frequency f in response to some sinusoidal sound or sinusoidal component of a sound. The speed with which the wave travels to the right along the basilar membrane depends on the frequency f. It also depends on the mass per unit length of the fluid above and below the membrane, on the mass per unit length of the membrane, and on the stiffness of the membrane. The cross section of the cochlea and the width, mass, and stiffness vary along the cochlea. A mathematical analysis shows how the speed of travel of a wave along the cochlea varies with distance. How the speed varies with distance depends on the frequency of the wave.

For any particular frequency, the speed of travel of the wave decreases in going from the left, basal end to the right, apical end. For each particular frequency the speed falls to zero at some *place* along the cochlea. Just to the left of that place the oscillation caused by the wave is greatest.

The motion of a wave along the cochlea from left to right is indicated in the upper part of Figure 7-5. Parts A, B, and C show the same wave as it moves to the right. The dashed lines show the envelope of the wave, that is, the greatest movement of the basilar membrane up or down at each point as the wave travels past that point.

Parts D through G of Figure 7-5 show the envelopes of four different waves along the basilar membrane for four different frequencies. We see that for low frequencies the place of greatest motion is closer to the apical end of the cochlea; for high frequencies the greatest motion is closer to the basal end. That is, the greatest vibration of the basilar membrane in response to a sinusoidal sound occurs at a particular place along it that depends on the frequency of that sound. The vibrations of the basilar

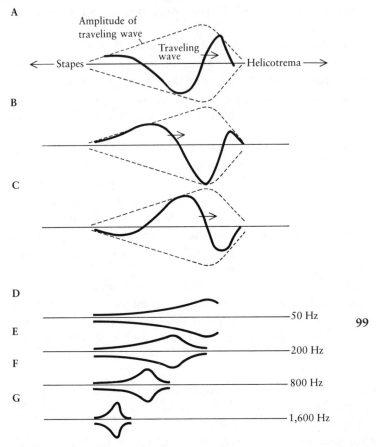

Figure 7-5 Waves traveling along the basilar membrane from left to right: **A**, **B**, and **C** show the same wave as it moves toward the right; **D**, **E**, **F**, and **G** show the envelopes for waves of four different frequencies. The drawings in this figure are adapted from the work of Nobel laureate Georg von Békésy, who made measurements on the cochleas of cadavers. Subsequent work on living animals has shown that the response of the basilar membrane is narrower and falls off very sharply to the right (toward the apical end). Békésy's drawings are used today chiefly because no other simple, cogent illustrations are available.

membrane excite electric pulses in nerve fibers that end on hair cells at this place on the basilar membrane (see Figure 7-6).

The reader may have seen by now that the frequency-dependent place at which the motion of the basilar membrane is greatest provides a means for sensing the frequency content of a sound wave, and hence, the pitch

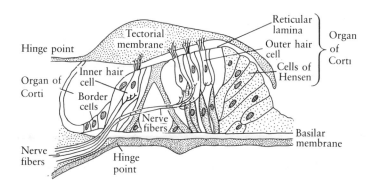

Figure 7-6 The basilar membrane and the structures surrounding it. As the basilar membrane moves, the hair cells send electric impulses along nerve fibers to the brain.

(and the timbre as well). Sine waves of different frequencies send messages to the brain along different nerve fibers; so the brain may judge the pitch of a wave by "knowing" which particular fibers carry the message to the brain. This is called the *place theory* of pitch perception, first put forth by Hermann von Helmholtz.

There is something more, however. A sharp pulse (a click) sounds just the same whether the pulse is a brief increase in pressure that pushes the eardrum in, or a brief decrease in pressure that pulls the eardrum out. When pulses or clicks follow one another in rapid succession, the ear can sense the rate at which the pulses arrive (if it is not too fast) and can infer a pitch from this rate.

As James L. Flanagan and Newman Guttman showed in 1960, sequences of pulses can be constructed for which the clue to pitch conveyed by the pulse rate conflicts with the spectral information we would get through a frequency analysis of the pulse trains. If the pulse rate predominates we infer one pitch; if the frequency analysis by the ear predominates, we infer a different pitch. If both pulse rate and frequency analysis are effective, there can be a pitch ambiguity, or there may be no clear pitch.

Consider the two pulse trains shown in Figure 7-7. The pulse rate is the same for both pulse sequence A and pulse sequence B. In A all pulses are positive, and the period (of repetition) is the time interval between pulses. In B the pulses are alternately positive and negative, and the period of repetition is twice the time interval between pulses. This means that the fundamental frequency of A, which is shown as a sine wave, is twice the fundamental frequency of B, also shown as a sine wave.

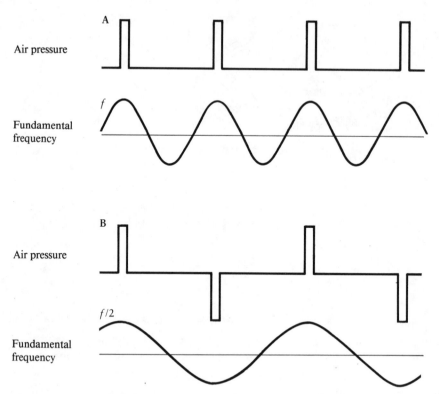

Figure 7-7 Two different pulse trains of sound. In part **A**, successive pulses represent sharp increases of air pressure. In part **B**, alternate pulses are positive and negative, that is, represent sharp increases and decreases of air pressure. At 100 pulses per second, both pulse trains sound the same pitch, but at 200 pulses per second, **B** sounds an octave lower than pulse train **A**.

What do we hear when we compare two such pulse trains? When the pulse rate is low enough, say, 100 pulses a second, pulse train B will sound like pulse train A for the same number of pulses per second, despite the fact that the fundamental frequency of B is only half that of A. But if the pulse rate of A is higher, say, 200 pulses a second, pulse train B sounds most like pulse train A when the two fundamental frequencies are the same, that is, when pulse train B has 400 pulses per second. The "match" is clearly of fundamental frequencies rather than of the number of pulses per second. At intermediate pulse rates time clues and frequency cues conflict, and pulse train B doesn't sound like A for any pulse rate.

How is it that at 100 pulses per second the ear doesn't make any effective frequency analysis of the pulse sequences? The ear can't. Whatever happens in the mechanism of hearing as a result of one pulse has died away before the next pulse comes along. At low rates the ear has no way of comparing the signs of successive pulses, and hence, no way of deducing either the fundamental frequency of the train of pulses or what harmonics of the fundamental frequency are present.

The response of the mechanism of hearing at low frequencies is more sluggish, and dies out more slowly, than the response to higher frequencies. Suppose that we filter out the lower harmonics of pulse patterns A and B, that is, those salient partials that most strongly convey the sense of pitch of musical tones. If we do this, the response to pulse rate rather than fundamental frequency can be pushed up to higher rates. Through such filtering the pulses are turned into short bursts of tone, of sinelike oscillation.

When pitch can be inferred from either repetition rate or spectrum, the clue given by spectrum tends to dominate. It seems that the first six or so partials give a very clear sense of musical pitch. When only high harmonics are present the tone tends to be buzzy and unmusical, and the sensation of musical pitch can be weak.

We have seen that there can be a conflict between time information and spectral information, with spectral information "winning" at comparatively moderate rates. But other experiments show that time information is preserved in nerve pathways at high rates. Suppose that we send a sine wave of constant frequency to one ear, and send the other ear a sine wave of the same frequency but of different phase. The sound that we hear will seem to be inside the head, but closer to the ear in which the sine wave peaks first. This effect persists for frequencies as high as 1,000 to 1,500 Hz. When a sine wave of one frequency goes to one ear and a sine wave of a slightly different frequency goes to the other ear, most of us hear *binaural beats*. These persist up to frequencies of 1,300 to 1,500 Hz. Animal experiments in which electric nerve impulses are observed by means of very fine electrodes inserted into individual nerve fibers show that time information is present for frequencies up to 4,000 to 5,000 Hz, but it is not clear that animals actually use such high-frequency time information.

It is clear that both time and place (of maximum motion of the basilar membrane) are available to us in judging periodic sounds. For nonperiodic sounds, we must rely on place information only. When we hiss by blowing air past our tongues, we can produce a sound of lower or higher pitch by moving the tongue back or forward. A hissing sound is not periodic, and any sense of pitchiness or brightness of a nonperiodic hiss must be judged by the place or places of vibration along the basilar membrane.

The place mechanism is also important in perceiving the timbre or quality of sounds. Suppose that in singing the same pitch we sing successively the vowel *u* as in *blue* and the vowel *e* as in *he*. We hear the pitch of the two vowels as identical, and this pitch agrees with the frequency of the first partial of the note sung. However, *e* sounds shriller than *u*. And we can distinguish vowels in pitchless, whispered speech.

We distinguish between vowel sounds by means of certain *formants*, or frequency regions of high energy with particularly strong overtones. These formants are centered on the resonant frequencies of the vocal tract and are changed by changes in the shape of the tract, depending on what vowels we utter. The frequencies of the formants all lie above the first partial of the sound we utter. The formants do not change frequency as we change the pitch with which we speak or sing a vowel. (This isn't quite true for sopranos; the timbre, or quality, of the sound changes when they sing high pitches, because they are singing above the formant range for a given vowel.) We must perceive the formant frequencies, and hence distinguish among vowels, by means of the place mechanism of the ear.

It is by means of the place mechanism that we distinguish the various frequency components in a sound. The critical bandwidth discussed in Chapter 6 should correspond to the "frequency range" along the basilar membrane that a single sine wave excites strongly. If a second sine wave of different frequency is to be perceived distinctly and independently, the excitation caused by a second sine wave should lie in a nonoverlapping region along the basilar membrane. That is, the second sine wave should be separated from the first by a critical bandwidth or more. The *location* of place of excitation of a sine wave along the basilar membrane, as in the judgment of pitch, is a different matter.

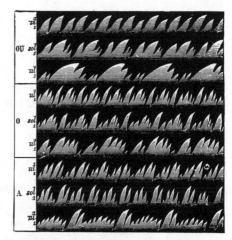

Figure 7-8 In a nineteenth-century experiment, it was found that vowel sounds would consistently impose the same pattern on the flame seen in the revolving mirrored drum regardless of the pitch level at which they were sung.

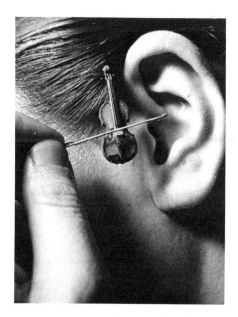

Figure 7-9   The world's smallest violin is quite shrill.

When we turn a sine wave on suddenly, at first the excitation along the basilar membrane is only roughly around the place corresponding to the wave's frequency. After some tens of cycles, a stable pattern of excitation develops. The amplitude of this stable excitation pattern falls very rapidly to zero as we move toward the apical end of the membrane. For musical tones made up of many harmonic partials, this should happen for the first six partials. The establishment of such sharp discontinuities might be the means through which the ear "measures" the places of the lower, salient harmonic partials that are essential to the sensation of musical pitch.

Georg von Békésy, who received a Nobel Prize for his work on the ear and hearing, made experiments with cochleas excised from human cadavers. He found the envelopes of the displacement of the basilar membrane in response to a sine wave to be rather broad, a shown in Figure 7-5. Later experiments he made with living animals indicated narrower envelopes, which broaden quickly after the animal dies. It appears that in the living cochlea some mechanism adds coherent energy to the oscillations set up on the basilar membrane and makes the frequency analysis sharper than it would otherwise be.

What we do know is that, in judging the periodicity and quality of sounds, the ear supplies the brain with two types of information. One type is information about time of occurrence and rate of repetition. Small differences in rate of repetition can be sensed from rates of one or fewer short pulses per second up to more than 100 pulses per second. Information about the frequency spectrum can be sensed from the lowest tone of the piano keyboard (which has a fundamental frequency of 27.5 Hz) far up beyond the highest — but the lowest tones are heard chiefly through salient lower harmonics, rather than through the fundamental and lowest harmonics.

We have seen through the examples of the patterns of pulses shown in Figure 7-7 that time information may be at odds with frequency information: At low rates a "match" between the different sequences of pulses is based on number of pulses per second; at higher rates a "match" is based on fundamental frequency.

Somehow, from clues of both time and place we derive our sense of both the pitches and the qualities of sounds. And, by comparing the information coming from the two ears, we somehow find out what direction the sound is coming from.

$\mathcal{8}$ *Power and Loudness*

e have all heard tales of the singer who shatters a glass with his (or her) voice. Such shattering is a purely physical phenomenon that depends on the power of the sound and the physical properties of the glass. Sounds can indeed be very intense. They can shake a wall. They can shake us. Yet like pitch, loudness is a sensation. We hear an intense sound wave as being louder than a feeble sound wave. But no sound wave is loud to a deaf person.

Unlike loudness, for most musical sounds, pitch depends on just one property of the sound wave: its periodicity. The relation between the power of a sound and its loudness is much more complicated.

We identify a shout as a loud sound even when we hear it in the distance. A shout is *different* from a normal tone of voice. We know that shouting requires more physical effort than speaking. A shout sounds raucous. It has more components of high frequency, and these vary with time in a distinctive way. So it is with musical instruments. Playing them loudly conveys a sense of force, of physical effort. Yet the actual sound power produced is important, too. In a quiet room we can understand a whisper. In a noisy room we must raise our voices in order to be understood.

What we can most easily measure in connection with loudness is sound power. We can study the loudness of sine waves (pure tones), which aren't really musical sounds at all. What we learn can guide us in considering loudness in music.

A sound source, such as a person talking, a trombone, or a stereo speaker, radiates sound waves whose total power we can measure in watts,

115

the same measure that we use for electric power. The sound waves spread out in all directions. The measure of the power, or energy, of a sound wave that reaches our ears is the power density measured in watts per square meter. This is the *intensity* of the sound wave. Instead of specifying the power density itself, it is customary to specify how many times the actual power is greater than the power of a standard *reference-level* sound wave with a power of 10^{-12} watt per square meter.* This reference level of power density is about the weakest sound that we can hear. The intensity of a sound above this reference level is the *intensity level.*

Most musical sounds have intensities that are millions of times the power of the reference level of sound intensity. Partly because it is difficult to write, read, or talk about huge numbers, and partly for other reasons, power ratios and intensity ratios are expressed in terms of *decibels*, abbreviated as dB. A certain number of decibels specifies a certain ratio of powers or intensities (power densities), as shown in Table 8-1. The dB scale is logarithmic, increasing in orders of magnitude. A sound with an intensity of 20 dB is *ten times* as great in intensity as a sound with an intensity of 10 dB (not twice as great). A sound of 40 dB intensity has an intensity a hundred times that of a sound with an intensity of 20 dB. We add or subtract anything measured in dBs; we do not multiply numbers of dBs.

In further discussions sound intensities are specified as a number of dB above reference level. *Sound level* as measured by a *sound-level meter* (see Figure 8-1) is measured in dB above reference level. We speak of *noise level* regarding environmental noise. The reading of a sound-level meter doesn't give the total power of sounds of all frequencies. Sound waves of very low and extremely high frequencies we simply don't hear, and a sound level meter shouldn't respond to them. As will be seen in this chapter, the ear is more sensitive to some frequencies than to others. The sound-level meter *weights* the powers or intensities of the frequency components of a sound wave before adding them to give an overall reading. Figure 8-1 shows three standard weighting curves. Curve C weights audible sounds nearly equally. Curve A corresponds to the sensitivity of the ear at moderate sound levels. Curve B is in-between.

Although we don't hear sounds of extremely low or extremely high frequencies, our ears are extremely sensitive. Suppose that we calculate the greatest distance at which we could just hear a 1-watt, 3,500-Hz sound source, the distance at which the sound will be at the threshold of our hearing, which is close to reference level. If we assume that the sound spreads out equally in all directions, the answer we get is 564 kilometers

---

*Note that 10^{12} means one followed by 12 zeros, but 10^{-12} means the reciprocal of this number, that is, one divided by 10^{12}, or 0.000000000001.

Table 8-1 Power Ratios and Decibels

P/P_r[a]	Ratio Expressed in Decibels[b]
1,000,000	60 dB
10,000	40 dB
1,000	30 dB
100	20 dB
10	10 dB
4	6 dB
1	0 dB
1/4 (=0.25)	−6 dB
1/10 (=0.1)	−10 dB
1/100 (=0.01)	−20 dB

[a] P is the power of the measured sound in watts per square meter; P_r is the reference-level power.

[b] Mathematically, the number of decibels is $10 \log_{10}(P/P_r)$.

(352 miles). Of course, there is no way of doing such an experiment, because there would be too much interference; but the calculation does give an idea of the remarkable sensitivity of our hearing.

Usually we can't hear sounds whose intensities are near reference level because the world around us is noisy. Early books tell us that the rustle of leaves in a gentle breeze produces a sound level of about 10 dB, that the sound level in a quiet garden in London can be as low as 20 dB, that in a quiet London street in evening with no traffic the sound level is about 30 dB, and that the night noises in a city are about 40 dB. It seems our world has grown noisier. Table 8-2 gives noise levels for a wide range of environments, and none is as quiet as some cited above. If you want quiet today you must go into a soundproof recording studio or concert hall. The average sound level in a concert hall with an attentive audience is about 40 dB, though the sound level in the hall when it is empty should not exceed 35 dB.

We should have such background sound levels in mind when we consider the intensity of musical sounds. An orchestra can produce a wide range of sound levels in a concert hall, from around 40 dB (the same as that of an attentive audience) to 100 dB (a million times as much power). For

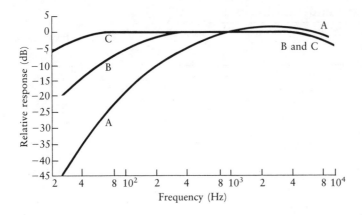

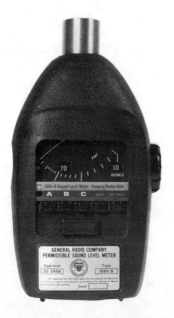

Figure 8-1 The intensity of sound or noise is measured with a sound-level meter (*bottom*). The meter does not ordinarily add the power of all frequency components of the sound and measure the sum. Rather, it "weights" the powers of different frequency components before adding. Weighting curve A (*top*) is commonly used, because the reading so obtained is closely related to loudness to the human ear for sounds of moderate intensity. Sound levels measured using the A setting (and weighting) of a sound-level meter are often quoted as dBA, as, for example, a noise level of 40 dBA. To measure the actual power of a sound, we would use the C weighting. The B weighting lies between A and C and would be appropriate for quite intense sounds.

Table 8-2 Noise Levels for Various Sources and Locations.

Source or Description of Noise	Noise Level (dBA)
Threshold of pain	130
Hammer blows on a steel plate (2 ft)	114
Riveter (35 ft)	97
Factories and shops	50–75
Busy street traffic	68
Ordinary conversation (3 ft)	65
Railroad station	55–65
Airport terminal	55–65
Stadiums	55
Large office	60–65
Factory office	60–63
Large store	50–60
Medium store	45–60
Restaurant and dining rooms	45–55
Medium office	45–55
Automobile at 50 mph	45–50
Garage	55
Small store	45–55
Hotel	42
Apartment	42
Home in large city	40
Home in the country	30
Motion picture theater, empty	25–35
Auditorium, empty	25–35
Concert hall, empty For full, add	25–35
Church, empty 5 to 15 dB	30
Classroom, empty	30
Broadcast studio, no audience	20–25

(Continued)

Table 8-2 Noise Levels for Various Sources and Locations. (*Continued*)

Source or Description of Noise	Noise Level (dBA)
Television studio, no audience	25–35
Television, studio, audience	30–40
Sound motion-picture stage	20–35
Recording studio	20–30
Average whisper	15–20
Quiet whisper (3 ft)	10–15
Threshold of hearing	0–5

comparison, the sound level of conversational speech ranges from 40 to 70 dB (a range of 30 dB, or a thousand times).

Table 8-3 gives the peak sound powers in watts for various musical instruments. It also gives the sound level at 3 meters from each instrument in open space, calculated by assuming that the sound travels equally in all directions, with no reflections. In an enclosed room, the sound level at a distance of 3 meters would be appreciably higher because of reflections from the walls.

If the sound of a musical instrument is not reflected or interrupted, its intensity drops 6 dB (that is, to a fourth of its value) every time we double

Table 8-3 Peak Sound Powers for Various Instruments, and Their Sound Level at a Distance of 3 Meters in the Open.

Instrument	Peak Power (watts)	Decibels Above Reference
Clarinet	0.05	86
Bass viol	0.16	92
Piano	0.27	94
Trumpet	0.31	94
Trombone	6.0	107
Bass drum	25.0	113

the distance from it. If the sound level is 90 dB at 3 meters from the instrument, it will be 84 dB at 6 meters, 78 dB at 12 meters, and 72 dB at 24 meters.

Sound Intensity and Loudness Level

As long as we stick to sound power or intensity, things are straightforward. We merely have to remember that sound level expresses the measured intensity relative to a reference intensity that is near the threshold of audibility, and that the ratio of intensities is expressed in decibels rather than as "times."

We have all observed that people differ in their hearing, as in other matters. Some have very acute hearing; some are quite deaf. The United States Public Health Service made an extensive survey of the *threshold of hearing* of many individual people; that is, of the level of sound at which they could just hear sinusoidal sounds of various frequencies. The results are shown in Figure 8-2.

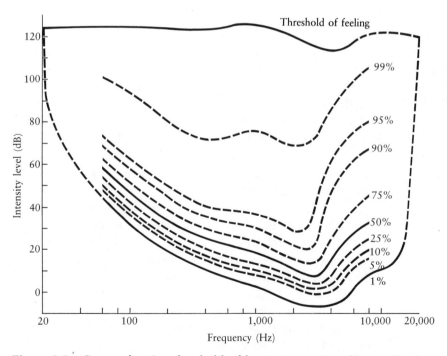

Figure 8-2 Curves showing threshold of hearing at various frequencies for a group of Americans: 1 percent of the group can hear any sound with an intensity above the 1 percent curve; 5 percent of the group can hear any sound with an intensity above the 5 percent curve; and so on.

In this figure, frequency of a steady tone is plotted from left to right, and sound intensity in decibels is plotted upward. The flat top curve shows the level of feeling, the intensity level above which sound hurts. We see that this is about 120 dB for all frequencies.

Each lower curve is designated by a percentage: 1 percent, 50 percent, ninety-nine percent. The 1 percent curve means that 1 percent of the subjects tested could hear a sound of a particular frequency whose intensity lay above the curve. For example, at 1,000 Hz, 1 percent of all individuals in the group tested could hear a pure tone whose intensity level was above 3 dB.

Clearly, how loud a sound is depends on who is listening. A sound that is inaudible to one person may be 10 or 20 dB above the lowest audible level for another.

Figure 8-2 tells us something else. Our ability to detect a sinusoidal sound, and sometimes the loudness of the sound, depends on its frequency, as well as on its intensity (power density).

All of us judge a particular sound as being louder if its intensity is increased. However, we can't discuss quantitatively the loudness of sounds for all people; so we must single out a group of persons, those with good or *acute* hearing, who are a minority among us.

For this group, loudness judged for sine waves is related to intensity and frequency by the curves shown in Figure 8-3. The curve at the bottom is for the threshold of hearing, the intensity level at which a sound of a particular frequency can just be heard. The other curves are *equal loudness curves*. That is, if two sinusoidal sounds of different frequencies have intensity levels lying on the same curve, they will sound equally loud.

We should note at once that the upper curves dip down less than the lower curves. For very loud sounds, the intensity level required to produce a given loudness doesn't change much with frequency. For very weak sounds it changes a lot. Equivalently, the loudness of a weak sound of a given intensity changes greatly as we vary its frequency; the loudness of a strong sound doesn't change much as we vary its frequency.

A consequence of this fact is that, as we turn the volume control of a stereo, the *relative* loudness of sounds of various frequencies changes. Some stereos have internal networks provided to compensate for this; mine has a switch labeled *loudness contour*. This is to be used when the sound produced by the speakers is less intense that the sound of the original music. When the loudness contour switch is on, the amplification is boosted at low and high frequencies to compensate for the dip in the low-intensity constant-loudness curves shown in the lower figure on the preceding page.

The equal-loudness curves in the figure are labeled in numbers of phons—20 phons, 40 phons, and so on. These curves can also be called

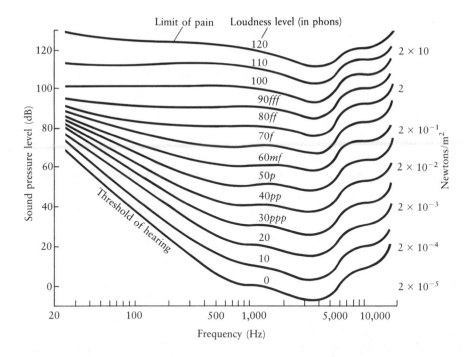

Figure 8-3 Constant-loudness curves for persons with acute hearing. All sinusoidal sounds whose levels lie on a single curve (an *isophon*) are equally loud. A particular loudness-level curve is designated as a loudness level of some number of *phons*. The number of phons is equal to the number of decibels only at the frequency 1,000 Hz.

isophons. A phon measures *loudness level.* The number of phons (of any particular equal-loudness contour) is merely the sound pressure level in decibels of an equally loud tone at 1,000 Hz. In other words, only at the frequency of 1,000 Hz does the number of phons equal the number of decibels — 30 phons = 30 dB, 60 phons = 60 dB, and so on. For all other frequencies, the relationship of phons to decibels has to be arrived at by experiment.

Loudness

Loudness itself is measured in *sones.* A sound with a loudness of 20 sones sounds twice as loud as a sound with a loudness of 10 sones. A sound with a loudness of 50 sones sounds twice as loud as a sound with a loudness of 25 sones. To measure phons relative to sones, we ask a person to turn up the intensity level of a sinusoidal sound until he or she hears the sound as being twice as loud. Surprisingly, we get consistent results.

Figure 8-4 shows the accepted relation between loudness in sones and loudness level in phons, which is the label of the equal-loudness curves in Figure 8-3.

We have noted that orchestral sound covers a range of intensity levels from 40 to 100 dB, approximately its loudness-level range in phons. This corresponds to a range in sones from about 1 to 50, or to between five and six doublings of loudness.

In the laboratory we can detect very small changes in loudness, but this is largely irrelevant to music. What is important is what difference we take into account in listening to music. These may or may not be closely related to *pianissimo (pp), piano(p), messo piano (mp), messo forte (mf), forte (f),* and *fortissimo (ff)*. This amounts to six levels of loudness, which is about the same as the number of doublings of loudness between the weakest orchestral sound and the strongest (as in Table 8-4). Is this coincidental? I don't think so. In music, the differences in dynamics, or loudness, that we specify correspond *roughly* to what a hearer regards as a doubling in loudness, that is, a doubling of loudness measured in sones; this accords with the markings from *ppp* to *fff* in Figure 8-3. Of course, performers don't measure dynamic levels in music numerically or absolutely; many subtle shadings lie interspersed among the dynamic markings given above.

The dynamic range of individual instruments other than the piano is of course much smaller than that of an entire orchestra, and for some wind instruments is exceedingly small. Figure 8-5 shows the relative intensities in decibels of the loudest and softest tones that expert instrumentalists could

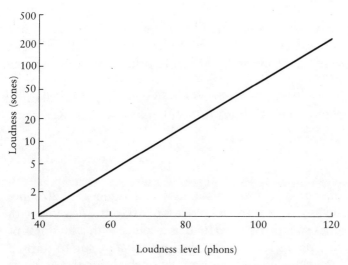

Figure 8-4 Relation between loudness in sones and loudness level in phons.

Table 8-4 Sound Intensities Expressed in Decibels.

	Intensity (W/m²)	Ratio I/I_0	Level (dB)
Threshold of feeling	$10^0 = 1$	10^{12}	120
fff	10^{-2}	10^{10}	100
f	10^{-4}	10^8	80
p	10^{-6}	10^6	60
ppp	10^{-8}	10^4	40
Threshold of hearing	10^{-12}	1	0

produce at various pitches. The size of the orchestra has increased dramatically over the past two hundred years. As composers wrote music with greater extremes of dynamic range, they wanted to attain a wide range of sound intensity despite the limitations of individual instruments.

Loudness of Combined Sounds

We should note that the curves in Figure 8-5 underestimate our judgments of the range of loudness of instruments played loudly and softly. As we have seen, a shout is different from a normal tone of voice both in spectrum and in how the spectrum varies with time. Musical instruments played loudly and softly also have different spectra which vary differently with time. We judge loud and soft partly by sensing physical effort. But perhaps we also judge loudness by sensing, unconsciously, a wider or narrower range of frequencies.

If two sounds of equal loudness are separated by more than a critical bandwidth,* when they are sounded together they are twice as loud as when either is sounded separately. More generally, the loudnesses of such sounds as measured in sones are added together.

If several sounds lie within the same critical bandwidth, first we must add the *actual* intensities (not the intensity levels in decibels) to get the total intensity. Then we must express the intensity level of this total intensity in decibels above the reference level. Then we must use the graph in Figure 8-3 to get the loudness level of the combined sounds in phons.

---

*About a quarter of an octave (a minor third); see Chapter 6.

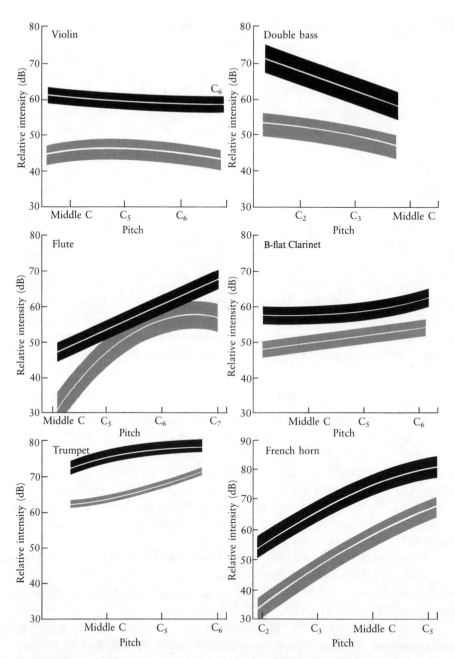

Figure 8-5 The dynamic ranges of various instruments. The broad upper bands show the relative intensities of the loudest tone that can be produced, as a function of pitch; the lower bands show the relative intensities of the softest tone that can be produced. The width of a band shows the range of intensities that may result when a player tries to produce "equally loud" tones.

Finally, we must use the graph in Figure 8-4 to get the overall loudness in sones.

We then find that there is a great difference between sounds separated by more than a critical bandwidth and sounds separated by less than a critical bandwidth. For the former, to double the loudness we need add only two sounds of equal loudness. For sounds lying in the same critical bandwidth, we must add eight equally intense sounds very close together in pitch in order to double the loudness.

Let us put this in another way. For simplicity, we will assume that the curves in the graph in Figure 8-3, which relate loudness level to sound intensity, are flat. For a single sine wave, we must multiply the power by 8 (an increase of 9 dB) to double the loudness. But we could also double the loudness by adding two equally intense partials, separated by at least a critical bandwidth (say, harmonic partials of frequencies f_0 and $2f_0$). That is, for a sine wave we have to increase the power eightfold in order to double the loudness; for two well-separated frequencies we need merely double the power to double the loudness.

We can put this still another way. Suppose that some given power of sound at a single frequency gives a particular loudness. We find that, if we divide this power between two frequencies, we increase the loudness by a factor of about 1.6. If we divide the power among the first six harmonic partials (f_0, $2f_0$, $3f_0$, $4f_0$, $5f_0$, $6f_0$), we increase the loudness by a factor of about 7.

Notoriously, single sine waves don't sound very loud. A group of harmonic partials of the same intensity level (power density) sounds much louder, partly because below 3,500 Hz a higher-frequency sound of a given intensity is louder than a lower-frequency sound of the same intensity (see Figure 8-3). To get a loud sound, it pays to put the power at frequencies higher than the fundamental, and separated by more than a critical bandwidth.

Successive harmonic partials above the sixth lie within a critical bandwidth. The sound of a harpsichord has many very high partials, and many of those lie within one critical bandwidth. That is why they sound rough or jangly. The sound would be even more jangly if the individual loudnesses of these high partials added, but they don't. Because successive high partials lie within the same critical bandwidth, the slow rise of loudness with intensity shown in the graph in Figure 8-4 helps protect us from jangle.

Summary

The following brief summary brings together the terms introduced in this chapter.

The total power of a source of sound is measured in watts, just as electric power is.

In a sound wave, we measure the *power density* in watts per square meter. This power density is called the *intensity* of the sound.

Intensity level expressed in decibels (dB) tells us the intensity of a sound relative to the intensity of a sound of *reference level*, which has a power density of 10^{-12} W per square meter, or 0 dB. The reference level is about the faintest sound we can hear.

The intensity level of a sound can be measured with a *sound-level meter*. If we want the power or intensity level, of a sinusoidal sound (pure tone) — or of any instrumental sound — to correspond to its actual power, we must set the sound-level meter to the flat or C scale, shown in Figure 8-1. But sounds with the same sound level may not sound equally loud.

Loudness level is a label for a constant-loudness curve such as one of the curves shown in Figure 8-2. The unit of loudness level is the *phon*. Suppose we choose the intensity level of the curve at 1,000 Hz as an example. A sinusoidal sound of 1,000 Hz and an intensity level of 70 dB has a loudness level of 70 phons. Any other sinusoidal sound of any other frequency whose intensity level lies on the same curve — that is, any sound of the same perceived loudness although of different frequency — has the same loudness level.

Pure tones with the same loudness level have the same loudness, but the loudness level doesn't tell us the loudness directly. The unit of loudness is the *sone*. If we double the loudness measured in sones, we hear the loudness as doubling. We can get from loudness level in phons to loudness in sones by using the graph in Figure 8-4.

I have saved one popular term for last. This is *sensation level*. Sensation level is the number of dB above the lowest intensity level at which a particular sound can be heard. That is, sensation level is the number of dB above the threshold of hearing for a sound of a particular frequency.

Suppose that an experimenter has a sound source of a certain frequency with a volume control that is accurately calibrated in dB, but is not calibrated relative to some known sound level. For two different settings of the volume control, the experimenter knows the ratio of the intensities (that is, the difference between the sound levels expressed in dB), but doesn't know the exact sound levels with respect to 10^{-12} watts per square meter, or 0 dB. The experimenter can produce a sound of 60 dB sensation level by first setting the volume control to produce a sound that the subject can barely hear and then turning the volume control up 60 dB.

Specifying sound levels as sensation levels rather than as intensity levels or sound levels (with respect to 10^{-12} watts per square meter, or 0 dB) is convenient for the experimenter. It is also sensible. The threshold of hearing varies with frequency, and it is different for different people. For the same person, it is a little different at different times. For many acoustical phenomena, including masking, scientists measure intensity relative to threshold, rather than to some arbitrary intensity.

9 Masking

We all know that weak sounds are drowned out by loud sounds. We might compare this to being unable to see in a blaze of extraneous light, but the ear is different from the eye. A bright light blinds us for a long time; the ear recovers very quickly. A bright light blinds us for all colors; a loud tone of a particular frequency renders sounds of only some other frequencies inaudible.

In 1876, Alfred Mayer published in the *Philosophical Magazine* a paper called "Researches in Hearing." Among other things, he excoriated conductors for obliterating the sounds of violins with the deeper and more intense sounds of lower-pitched wind instruments. He observed that an intense sound of low pitch can mask a weaker sound of high pitch, but a sound of high pitch cannot mask a sound of low pitch.

This fits well with our understanding of how waves travel along the basilar membrane, as described in Chapter 7. The place of greatest excitation of the basilar membrane for tones of low frequencies is toward the far or apical end of the cochlea, but that for tones of high frequency is toward the near or basal end of the cochlea. In traveling along the cochlea, the wave excited by a high-frequency tone will never reach the place of a low-frequency tone. But in reaching their place, the waves set up by low-frequency tones must travel past the places of all tones of higher frequency. We might expect that the excitation of the basilar membrane at these places could interfere with the perception of high-frequency tones. If the low-frequency tone is strong enough, it can indeed interfere with our hearing tones of higher frequency.

130

When a weak sound is obscured by a stronger sound, it is said to be *masked* by the stronger sound. The strong sound is called the *masker*. The weak sound that is masked is called the *maskee* or *signal*. Masking by a strong sound may be likened to an impairment of hearing. The masker, in effect, raises our threshold of hearing, that is, it raises the intensity that a sound must have for us to just hear it.

The first systematic investigation of masking was carried out at Bell Laboratories in 1924 by R. L. Wegel and C. E. Lane, and the results are available in Harvey Fletcher's *Speech and Hearing in Communication*. Figure 9-1 shows these results.

The figure presents results for masking tones of six different frequencies. Let us consider one set of curves, that for a masker frequency of 1,200 Hz (part D). The horizontal scale shows the frequency (in hertz) of the tone that is masked. The vertical scale shows *threshold shift* in dB, that is, the increase in intensity level of the maskee (above threshold in the *absence* of the masker) that will just enable us to hear the maskee in the presence of the masker. Five curves are shown for masker sensation levels of 20, 40, 60, 80, and 100 dB.

Still in part D, consider the topmost curve, that for a very loud masker with a sensation level of 100 dB. A masker will be heard if its sensation level—above threshold *in the absence of the masker*—lies above this curve. Consider two maskees of the same frequency, 1,600 Hz, one with sensation level (in the absence of the masker) of 80 dB, the other of 60 dB. The first maskee lies above the curve (reading along the threshold shift) and will be heard; the second lies below the curve and won't be heard. In effect, the 100-dB masker has raised the threshold of hearing for 1,600-Hz sounds by about 69 dB (again, along the threshold shift) above the threshold in the absence of the masker.

If we examine the curves, we see that there is no threshold shift for maskees with frequencies below 400 Hz. No matter how strong the 1,200-Hz masker, it causes no appreciable vibration of the basilar membrane at the 400-Hz place.

We note that the greater the intensity of the masker, the greater the upward shift of the threshold. We should expect this. As the intensity of a masker is steadily increased, a maskee of constant intensity must finally be masked.

We can also note that 1,200-Hz maskers with sensation levels of 20 and 40 dB produce no masking at frequencies above 2,400 Hz. Though the waves excited on the basilar membrane by such weak maskers pass the 2,400-Hz place on their way to the 1,200-Hz place, they don't interfere with the perception of 2,400-Hz tones.

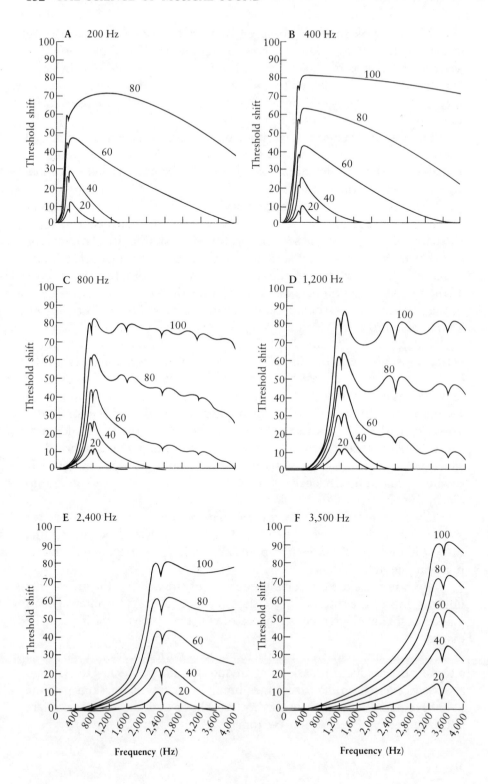

For maskers of higher sensation levels, this is not so, and the shape of the curves for these higher-level maskers tells us something. Let us look at the 100-dB curve in part D, for example. We see little sharp dips at frequencies of 1,200 Hz, 2,400 Hz, and 3,600 Hz. *Very close to particular frequencies, the masker produces less masking than it does at nearby frequencies.* Why is this?

Consider masking by a 1,200-Hz masker of a tone of 1,220 Hz. We hear 20-Hz beats between the two tones. In essence, the combined amplitude of any two sinusoidal tones of slightly different frequencies rises and falls with time. Sometimes the pressures of the masker and the maskee peak at the same time; a little later the pressure of the maskee is least when the pressure of the masker is greatest, and the combined pressure is lower than average. Still later the pressures add again. This rise and fall of the sum of the pressures, this beat between the two tones, repeats regularly at a rate of 20 Hz, the difference in frequency between the masker and the maskee. This beating enables us to detect the presence of a weak maskee even when we can't hear it as a tone of different frequency. Hence beats effectively lower the threshold shift and decrease the amount of masking.

This explains why there is a dip in the masking curve at 1,200 Hz, the frequency of the masker, but why are there dips at 2,400 and 3,600 Hz? Nonlinearities (failures to produce an output signal that is a faithful reproduction of the input signal) in either the electronic equipment used or in the middle or inner ear could cause a very strong 1,200-Hz masker to produce frequency components whose frequencies are harmonics of the 1,200-Hz masker, such as a 2,400-Hz second harmonic and a 3,600-Hz third harmonic. These waves of harmonic frequency would produce vibrations of the basilar membrane at the 2,400-Hz place and the 3,600-Hz place; these vibrations would beat with the maskee tones of frequencies near 2,400 Hz and 3,600 Hz and lower the masking curve.

Nonlinearities in the middle and inner ear could account for the fact that very intense maskers produce large threshold shifts for sounds whose

Figure 9-1 (*Facing page*) Masking curves. In each box, the frequency of the tone that does the masking is given at the top; the horizontal scale gives the frequency of the tone that is masked; and the vertical scale gives the amount of masking in decibels, that is, how much more intense than threshold (in absence of masking) the masked tone must be in order to be heard in the presence of the masker. Except for the masker frequency of 200 Hz (part **A**) there are five curves for masker sensation levels of 20, 40, 60, 80, and 100 dB. For a given curve (level of masker), the maskee will be heard if its sensation level in the absence of the masker lies above that curve, and will not be heard if its sensation level lies below the curve.

frequencies lie above the frequency of the masker. As a wave of very large amplitude and low frequency, and waves of its harmonic frequencies, reached higher frequency places on the basilar membrane, they could interfere with our ability to hear weak, high-frequency sounds.

If we look at the masking curves for low-level maskers, we find that masking extends for only a narrow range of frequencies. We might expect this frequency range to correspond to the critical bandwidth. In fact, the critical bandwidth for masking is pretty much the same as the critical bandwidth for dissonance and the critical bandwidth governing loudness.

Masking by Noise

So far we have discussed masking by single-frequency sinusoidal sounds, that is, by pure tones. Noise made up of a broad band of frequencies also produces masking. A colleague and I once shared an office that looked out on the West Side Highway in lower Manhattan. In summer, when the windows were open, streams of passing cars flooded the office with a broad-band noise, an intense sound with a continuous range of frequencies. I soon got used to the noise, and found that it gave me complete privacy. I couldn't hear a visitor talking to my colleague who sat at a desk behind me. I couldn't even hear his visitor enter or leave the room.

Many masking experiments are carried out using noise rather than pure tones. Figure 9-2 compares, for the same subject, masking by a pure tone of 400 Hz (part A) and masking by a narrow band of noise extending from 365 Hz to 455 Hz (part B). We should note that in this figure the levels of the curve are *intensity levels* with (respect to 10^{-12} W per square meter, or 0 dB), not sensation levels as shown in Figure 9-1. We see from Figure 8-3 (Chapter 8) that at 400 Hz the threshold of hearing is at an intensity level of 10 dB; so the curve marked 60 dB in both parts of the figure below would correspond to a curve marked 50 dB (sensation level) in Figure 9-1.

Figure 9-2 (*Opposite page*) Masking for a pure-tone masker (**A**) compared with masking for a 90-Hz band of noise (**B**). In part **A**, the masker frequency is 400 Hz; in part **B**, the masker (noise) is centered on 410 Hz. In these curves the labels on the curves specify sound level. The threshold of hearing at 400 Hz is about 10-dB intensity level; so the sensation level of the maskers for these curves is about 10 dB less than the intensity level, the labels shown on the curves. Narrow-band noise usually masks more than a pure tone, but for the highest curves the pure tone masks more at higher frequencies.

A

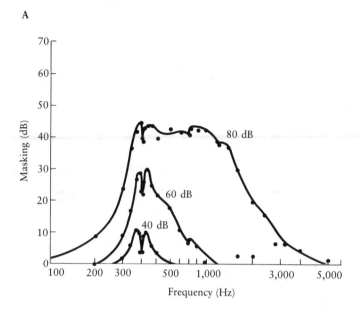

B

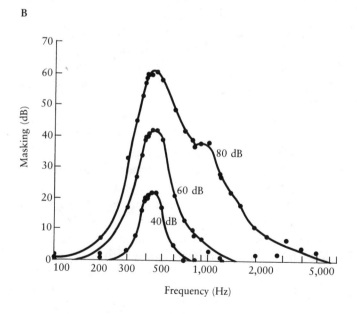

In comparing masking by noise (part B) with masking by a pure tone (part A), we find that in masking by noise there are no dips due to beats (unless the slight dip at 800 Hz for the 80 dB curve is such an effect). We hear noise plus a pure tone simply as noise plus a pure tone.

The figure shows that, at most frequencies and levels, noise masks more effectively than does a pure tone of the same intensity. But the masking of a pure tone of intensity level 80 dB falls off less rapidly with increasing frequency than does masking by a narrow-band noise of the same intensity level.

Regarding masking by wide-band noise, for moderate levels, the only portion of the power of wide-band noise that is effective in masking a pure tone is the power of those frequency components whose frequencies lie within a critical bandwidth centered on the pure tone. For a particular kind of broad-band noise called *white noise*, the power density is independent of frequency; that is why the noise is called white. The power or intensity of the part of the noise that lies within any band of frequencies is proportional to the bandwidth. If the critical bandwidth is proportional to frequency (say, a quarter octave), the effective intensity of the white noise in masking a pure tone will be proportional to the frequency of the tone.

Masking by white noise is important in psychoacoustic studies. In the perception of music, masking by other sounds is much more important. We noted in Chapter 8 that the intensity level in an auditorium with an attentive audience is around 40 dB. What does this mean to listeners? What level of musical sound will such audience noise mask? How can we find out? We don't have masking data for audience noise and musical instruments or orchestras. We must infer what we can from masking data for pure tones and for noise.

Figure 9-1 gives data for masking of a tone by a tone. We see from the figure that, for low masking levels, we can hear tones some 10 to 15 dB below the masker level. But audience noise is not a pure tone.

If we consult part B of Figure 9-2, which gives masking of a tone by noise, we conclude that we can just hear a tone if it is some 7 dB below the level of *narrow-band* masking noise. (Remember that for this figure a masker level of 40 dB corresponds to a sensation level of 30 dB.) Other data on masking of tones by noise indicate that a tone becomes inaudible at from 2 to 6 dB below the noise in a critical bandwidth.

Masking of Noise by a Tone

We have discussed the masking of a tone by a tone, and a tone by noise. Masking of narrow-band noise by a tone is very different. We tend to hear the noise despite the presence of the tone. In effect, the noise tends to make the tone wavering and noisy. Indeed, a critical bandwidth of noise can be masked completely by a tone in the center of the band only when the tone is about 24 dB stronger than the noise.

We conclude that, if musical sound had the same spectrum as audience noise, we could hear it if its level were a few dB below that of audience noise, perhaps in the range from 34 to 38 dB for an audience-noise level of 40 dB. But we could hear a pure tone with a much lower intensity level, for only that part of the audience noise within a critical bandwidth around the pure tone would mask it effectively. Saying that the noise produced by an attentive audience has an intensity level of 40 dB may give the false impression that such audience noise is quite annoying. This amount of noise won't much interfere either with many instruments playing together or even with single instruments played softly.

Perhaps it is more pertinent to ask how singers manage to be heard above an orchestra. Johan Sundberg has found the explanation by comparing the frequency distribution of power (the sound power or intensity spectrum) for an orchestra and for a tenor (in this case, the late Jussi Björling). Figure 9-3 shows what he found. The intensity level of orchestral sound falls off rapidly at frequencies above 500 Hz. The power of the

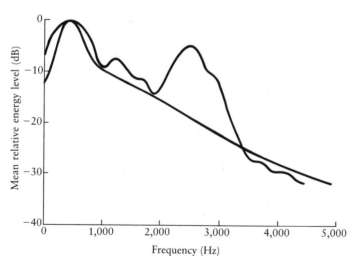

Figure 9-3 Comparison of the frequency distribution of sound power for an orchestra (smooth) and for a tenor (wiggly). The curves are adjusted to have the same peak values. Orchestral sound falls off steadily in power as the frequency is increased above 500 Hz. The singer's voice has a second peak of power between 2,000 and 3,000 Hz, which gives him an advantage against masking of about 13 dB (or 20 times) in this frequency range.

singer's voice peaks at about 500 Hz, but it peaks again at about 2,500 Hz, which gives the singer an advantage over the orchestra of about 13 dB (or 20 times in intensity). In this frequency range, the singer can equal the orchestra even if he produces only a twentieth as much sound power.

The Cocktail-Party Effect

Have you ever noticed that in a crowded room you can hear what someone nearby is saying, despite the babble of other speakers who are no farther away? This ability to distinguish one voice among many is called the *cocktail-party effect*. It can be effective in sorting out the instrumental sounds of a chamber group if they are playing while you are nearby.

We might think that the cocktail-party effect is merely a matter of "paying attention." No doubt paying attention is important, but there is more to the cocktail-party effect than that. The effect vanishes if we listen over a single audio channel. In Chapter 7, it was noted that time information is preserved in the auditory pathways. Imagine that the speaker that we wish to hear is dead ahead, whereas the noise that we wish to ignore comes from the side. The sound of the speaker will reach both ears at the same time and in phase; the noise won't. We can take advantage of this in listening, so that somehow the neural speech signals add in phase and reinforce one another (in our *binaural* hearing), whereas the noise signals don't add in phase and may partially cancel. (The mechanism of the cocktail-party effect is really more complex than this and is not fully understood. It enables us to single out a voice coming from a direction other than dead ahead.)

The cocktail-party effect is extremely important in listening to *nearby* sounds, and it may help somewhat in ignoring the coughs of near neighbors while listening to music in a large hall. But for concert audiences the phenomenon of masking, without regard to binaural effects, is predominant. It is also very important in regard to distortion in sound systems used for the reproduction of music. Nonlinearity in sound systems produces frequency components that were not present in the original music. If these lie close to frequencies already present in the music, they will be masked and we won't notice them. But if they fall in frequency ranges where the intensity of the music itself is low, the frequencies produced by nonlinearities can be very annoying. We notice this particularly in small rattles or in the noise produced when a misaligned speaker coil touches the magnet around it and produces a faint, high-frequency rustling sound as it moves. The power of this rustling sound is very small compared with the power of the music itself, but we hear the resulting sound clearly because it is not masked by the lower-frequency sounds of the music.

Like pitch and loudness, masking is of great importance in the perception of musical sound. Masking is important in the reproduction of sound, for unmasked distortion grates on the ear; masked distortion does not. Masking is important in listening to sound. In a noisy automobile we can't hear the soft passages in a composition, but in a quiet concert hall we can.

Perhaps the most fundamental importance of the masking of musical sounds lies in the fact that one musical sound can mask another. Intended subtleties will be obscurities if the sounds intended to produce them are masked by other sounds.

10   *Other Phenomena of Hearing*

*W*hy do psychoacousticians single out and measure various phenomena of hearing? Partly to find measurements that give consistent results. Partly to accumulate data that will suggest, validate, or destroy various theories of hearing.

Many of the measurements made by psychoacousticians are of little practical value for music. For example, as we have seen, most musical tones have two quite different qualities: pitch, which depends on periodicity; and brightness, which depends on the relative strengths of the partials. Sine waves have only one frequency component, and in experiments done with sine waves, the naive listener might respond to either the sense of brightness or the sense of pitch. However, experimental data are our only guard against hasty conclusions and self-deceit. Although we would prefer to have more experimental data truly relevant to musical sounds, we should certainly not disregard what psychoacoustic data we do have.

The Just Noticeable Difference

One classical measurement of psychoacoustics is the *jnd* (just noticeable difference), or *limen*. The most common jnds are those of intensity and of frequency, measured for sine waves. Unhappily, measured jnds differ from one method of measurement to another, from one subject to another, and even for measurements taken on the same subject at different times.

In measuring jnds, some early experimenters "wobbled" the intensity or frequency, and found how small a wobble was just noticeable. One alternative is to alternate a tone fixed in intensity or pitch with an adjustable tone, and to ask the subject to match the adjustable tone to the fixed tone. The average error can be taken as the jnd. Most often today, successions of two tones of different intensities or pitches are presented to the subject. The subject is asked which of the two is higher in intensity or pitch. If the subject makes errors, the difference is increased; if the subject is correct, the difference is decreased. The jnd is taken as the difference for which the subject is correct 75 percent of the time.

The jnd of intensity depends on both frequency and intensity. Table 10-1 shows a classic measurement of the jnd of intensity in dB for sine waves of various frequencies and sensation levels. The jnd is least, about a quarter of a dB, at high levels and at frequencies from 1,000 to 4,000 Hz. At frequencies above 100 Hz and intensity levels above 40 dB, it does not exceed 1 dB.

The jnd of frequency also depends on frequency and intensity. Table 10-2 shows the jnd of frequency for pure tones of various frequencies and sensation levels. We should note that for sensation levels of 30 dB and above, the jnd is 10 cents or less (less than the mistuning of the thirds in the equal-tempered scale) for frequencies of 500 Hz and above. Below 500 Hz the jnd rises rapidly with decreasing frequency, and at 40 dB it is 70 cents (almost a semitone) at 31 Hz (the lowest B on the piano keyboard). Pitch perception is very poor for sinusoids of low pitch. For actual musical tones in the mid- and high range, the jnd of pitch is comparable to that for sine waves: In these ranges the fundamental frequency is important to pitch. But the jnd is smaller than for sine waves toward the low end of the piano keyboard.

Table 10-1 The Minimum Detectable Changes (jnd) of Intensity in Decibels for Sine Waves.

Frequency (Hz)	Sensation Level											
	5	10	20	30	40	50	60	70	80	90	100	110
35	9.3	7.8	4.3	1.8	1.8							
70	5.7	4.2	2.4	1.5	1.0	.75	.61	.57				
200	4.7	3.4	1.2	1.2	.86	.68	.53	.45	.41	.41		
1,000	3.0	2.3	1.5	1.0	.72	.53	.41	.33	.29	.29	.25	.25
4,000	2.5	1.7	0.97	0.68	.49	.41	.29	.25	.25	.21	.21	
8,000	4.0	2.8	1.5	.9	.68	.61	.53	.49	.45	.41		
10,000	4.7	3.3	1.7	1.1	.86	.75	.68	.61	.57			

Table 10-2 The Minimum Detectable Changes (jnd) of Frequency in Cents for Sine Waves.

Frequency (Hz)	Sensation Level										
	5	10	15	20	30	40	50	60	70	80	90
31	220	150	120	97	76	70					
62	120	120	94	85	80	74	61	60			
125	100	73	57	52	46	43	48	47			
250	61	37	27	22	19	18	17	17	17	17	
550	28	19	14	12	10	9	7	6	7		
1,000	16	11	8	7	6	6	6	6	5	5	4
2,000	14	6	5	4	3	3	3	3	3	3	
4,000	10	8	7	5	5	4	4	4	4		
8,000	11	9	8	7	6	5	4	4			
11,700	12	10	7	6	6	6	5				

Measurements of just noticeable differences of intensity for sine waves are not necessarily directly relevant to the intensity levels — as measured by a sound level meter — with which a skilled musician plays the successive tones of a scale at even loudness. In one early experiment even scales were played over the entire range of a violin, a B-flat trumpet, and a bass clarinet. Levels ranged irregularly but reproducibly over an intensity range of as much 10 dB. Sine waves of different pitches are much "alike." However, successive scale tones of an actual instrument are different in spectrum and its time evolution, and in loudness, for the same intensity (power).

The jnd of frequency seems more relevant to actual musical performance, although, as we have noted, jnds for musical sounds of low pitch are smaller than the jnds for sine waves of low pitch. Some early data on trios indicate a variability in pitch comparable to the error of the thirds in equal-tempered tuning. Later data have indicated deviations in frequency of about 10 cents for a good violinist, a figure not too far from the jnd for sine waves.

Perhaps jnds of frequency tell us how accurately a musician can tune an instrument. There was an electronic tuning test at the Bell System exhibit in Disneyland. By pressing buttons you could hear either a tone of fixed frequency or a tone whose frequency you could adjust, but not both at the same time (so you couldn't hear beats). After you had matched the frequencies as closely as possible, the machine scored your performance. My wife, a musician, did much better than I.

Other aspects of auditory perception are much more important to musical sound than are jnds of frequency or intensity. At the head of these I would put what is called the *precedence effect*, or the *Haas effect*.

The Precedence Effect

This effect you can demonstrate to yourself if you have a stereo system, and especially if you tune to a monophonic (nonstereo) channel, so that exactly the same sound comes from each speaker. If you stand equidistant from both speakers, you hear the sound as coming from a phantom source midway between the speakers. But if you stand a foot or more closer to one speaker than to the other, all the sound seems to come from the nearer speaker.

I once checked this in a car that had a stereo radio. In the exact middle of the front seat, I heard the announcer's voice as a compact sound source dead ahead, midway between the speakers. As I moved slightly to the right the sound source at first became diffuse. As I moved farther, all the sound clearly came from the right-hand speaker.

Like other stereo systems, the car stereo has a knob (a balance control) that changes the relative intensities of the sounds coming from the right and left speakers. When one is sitting in the driver's seat, for a monaural or centered signal (the announcer's voice, for example) all the sound seems to come from the left speaker unless the sound from the right speaker is made more intense than that from the left. Indeed, even for a stereo signal most of the sound seems to come from the left speaker unless the sound from the right speaker is made more intense. By making the sound from the right speaker more intense, we can hear sound from both speakers and get a stereo effect. However, adjusting the relative intensities doesn't cure the disparity between time of arrival (at our ears) from the left and right speakers. If the speakers are equidistant and the intensities are equal, we hear a speaker's voice as coming from a distinct direction, from a compact source. If the speakers aren't equidistant, and we increase the intensity of sound from the far speaker, we do hear some sound from it, but we no longer hear a speaker's voice as coming from a compact source. The source of the voice seems extended and fuzzy.

I have said that, if a sound reaches us with equal intensities from two sources, we hear all of it as coming from the nearer source if the difference in distance is about a foot or greater. There's a limit to this, of course. If the difference between the distances to the two speakers is great enough (about 20 to 25 meters, or 60 to 80 feet), you hear two sources. The speaker that is farther away produces a distinct echo of the sound from the nearer speaker. You can easily demonstrate this transition from fusion to echo as the American physicist Joseph Henry did in about 1849. Stand in front of an extensive, smooth wall, and clap your hands. If you're less than ten meters from the wall (sound then travels 20 meters in reaching the wall and returning), you hear a single sound. If you're farther than 13 meters or 40 feet away, you hear an echo of your handclap.

The precedence effect, the fact that a sound seems to come from the direction from which it reaches us first, is bad for stereo, but highly desirable in everyday life. When someone speaks to you in a hard-walled room, you hear all the sound as coming from his or her mouth, even though much of the sound that reaches you has been reflected from the walls, sometimes several times. This reflected sound adds to the loudness, but doesn't keep you from identifying the direction of the source. The same is therefore true for musical instruments. If you are close enough to a chamber group, you hear each sound coming from its proper direction, even though much of the total sound that you hear reaches you after reflection from the walls.

Reflections from the walls of a room do not confuse our sense of direction, but they add to the intensity of the sound, and they add to its quality as well. While I was in Paris, working at IRCAM, my wife practiced there on a Yamaha concert grand in a studio that had very sound-absorbing walls. She found this exasperating, because, as loud as she tried to play, she produced very little sound. Furthermore, that sound was very "dead," a matter to which we shall return.

Besides increasing the intensity of sounds, reflections from walls helps us judge the distance of a sound source. For the same intensity of sound reaching an observer, the distance from the source is judged to be greater if the same sound comes from several loudspeakers at different distances than if it comes from a single speaker. This is quite reasonable. If someone speaks to us in a room, we hear mostly direct sound if the person is close to us, mostly reflected or reverberant sound if he or she is far away. In an anechoic chamber (a room whose walls do not reflect sound), if you stand behind a person and whisper, the person will think that you are close, regardless of distance. John Chowning used varying amounts of reverberation to give a sense of varying distance to computer-generated sounds in his compositions *Stria* and *Turenus*.

Playing in a "dead" room can be exasperating to performers. Musicians who played in Philharmonic Hall (now Avery Fisher Hall) in Lincoln Center before extensive modifications were made found it difficult to hear themselves play, because the wall behind the performers was made of sound-absorbing material (see Table 10-3). I've been told that performers

Table 10-3 Absorption Coefficients of Some Building Materials.

Material	Frequency (Hz)					
	125	250	500	1,000	2,000	4,000
Marble or glazed tile	.01	.01	.01	.01	.02	.02
Concrete, unpainted	.01	.01	.01	.02	.02	.03
Asphalt tile on concrete	.02	.03	.03	.03	.03	.02
Heavy carpets on concrete	.02	.06	.14	.37	.60	.65
Heavy carpets on felt	.08	.27	.39	.34	.48	.63
Plate glass	.18	.06	.04	.03	.02	.02
Plaster on lath on studs	.30	.15	.10	.05	.04	.05
Acoustical plaster (1 in)	.25	.45	.78	.92	.89	.87
Plywood on studs (1/4 in)	.60	.30	.10	.09	.09	.09
Perforated cane fiber tile, cemented to concrete, 1/2 in thick	.14	.20	.76	.79	.58	.37
Perforated cane fiber tile, cemented to concrete, 1 in thick	.22	.47	.70	.77	.70	.48
Perforated cane fiber tile, 1 in thick, in metal frame supports	.48	.67	.61	.68	.75	.50

like to play about 15 or 20 feet in front of a reflecting surface. They can hear themselves play. This seems natural to them and helps them.

Reflected sound, or reverberation, is even more important to audiences. I once heard a recording of a pipe organ that had been made in the organ loft. It sounded like an electronic organ, because there were no reverberations. Reverberation blends and adds richness to sounds. My wife and I once heard a student orchestra practicing in Sainte-Chapelle in Paris.

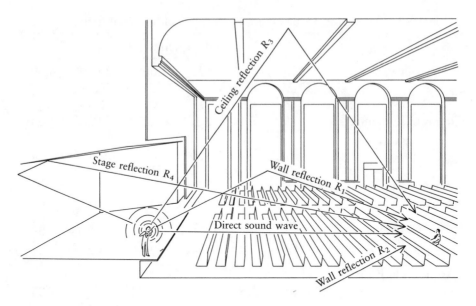

Figure 10-1 Reflections of sound.

They produced a shimmering cascade of sound, especially from the trumpets. "It sounds wonderful," I said. "But they aren't together," my wife replied. Indeed, they weren't, but the individual notes were so lost in grand sonorities of reverberation that I hadn't noticed.

Reverberation time is defined as the time it takes a sound to decrease to 60 dB below its initial intensity. Speech becomes hard to understand when the reverberation time is greater than one second and is clearer when the reverberation time is half a second. Two seconds is fine for music, and some modern organ music seems designed for cathedrals with still longer reverberation times (see Figure 11-2).

Echoes

Echoes can be annoying. As we have noted, when we hear the same sound from two sources 20 to 26 meters (60 to 80 feet) away (a time difference of 60 to 80 milliseconds), we hear an echo. We hear an echo in an auditorium if we get one strong, delayed reflection from a flat surface, such as the front of a balcony. But in a well-designed auditorium, single strong reflections are prevented. Instead, we get a multitude of reflections that reach our ears at different times and from different directions. Because the sound reaches our ears first by a direct path without reflections, we judge the whole of the sound to come from a small source on stage. However, its quality is very different from what it would be without reverberation.

Strong, single, distant reflections are undesirable. So are small nearby reflections. We are told to put loudspeakers either against or in walls, or else far away from them. The effects of close reflections are even more apparent with microphones. If you put a microphone 2 or 3 inches from the surface of a table, whatever it picks up, voice or music, sounds distorted. Sounds are "colored." We hear some frequencies as emphasized, some as suppressed.

We can easily see why this is so. Figure 10-2 shows a microphone in front of a perfectly reflecting wall. A sound coming from the right reaches it twice, once directly and once reflected, at a time

$$t = \frac{2L}{v} \text{ seconds}$$

later, where v is the velocity of sound.

Consider Figure 10-3. If the sound is a tone of frequency

$$f = \frac{v}{4L}$$

a trough of the reflected wave (in color) will reach the microphone just as a crest of the original wave reaches the microphone. The two waves, direct and reflected, will cancel; so the microphone won't pick up anything at that frequency. The response of the microphone to tones of increasing frequency will also go to zero at frequencies $v/4L$, $3v/4L$, $5v/4L$, and so forth, as shown in Figure 10-4.

We can prevent the unwanted effects of reflected sound by putting the microphone very close to the wall. We could prevent gross coloration

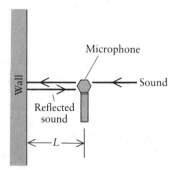

Figure 10-2 A microphone in front of a perfectly reflecting wall. Sound reaches the microphone with equal intensity, both directly and after being reflected from the wall.

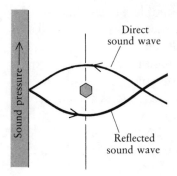

Figure 10-3 Direct and reflected sound waves for the microphone in Figure 10-2. If the frequency of the sound is $v/4L$, as shown, the pressure of the reflected sound wave will be equal and opposite to the pressure of the direct wave; so the microphone will pick up nothing.

of sound by putting the microphone far from the wall, though the sound will be affected by the reverberation of the room.

We tend to hear reverberation rather than coloration when the difference between frequencies at which the response goes to zero or dips sharply is small compared with a critical bandwidth (roughly a minor third, about 1/5 of the frequency in question). The frequency difference between successive dips in the response curve in Figure 10-4 is $v/2L$. Hence, for reflections not to distort the sound, we should have

$$\frac{v}{2L} \text{ smaller than } (1/5)f$$

or

$$L \text{ larger than } (5/2)(v/f)$$

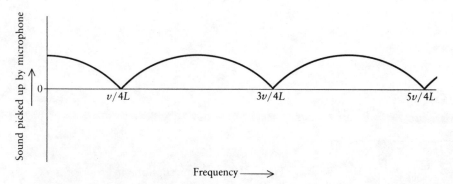

Figure 10-4 How the sound picked up by the microphone in Figure 10-2 varies with the frequency of the sound wave.

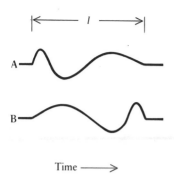

Figure 10-5 Two sound waves (**A** and **B**) have different waveforms but the same energy spectrum. Measurements made by David M. Green show that such waves are heard as different if their duration is greater than 2 milliseconds. Here wave **B** is wave **A** reversed in time.

If we want to prevent distortions for frequencies down to 100 Hz, we must make L at least 9 meters. (Actually, the critical bandwidth is greater at lower frequencies; so 9 meters is larger than needed.) Anyway, you can now see why you must keep your microphone well away from floors and walls when you are recording music.

One other matter concerning the perception of music is worth mentioning: our ability to distinguish small periods of time. In 1973 David M. Green published some interesting results on temporal acuity. He measured the ear's ability to discriminate between two signals that have different waveforms but the same energy spectrum. An example of such signals is any short waveform and the same waveform reversed in time, such as those in Figure 10-5. Here part A is a sound of decreasing frequency, part B of increasing frequency. Green found that the ear can tell the difference between two such waveforms if their duration is greater than 2 milliseconds.

The phenomena of the precedence effect and of echoes are of extreme importance to music. If one of two sources of the same sound lags behind the other by more than a millisecond (or foot), the earlier source "swallows up" the later as far as our sense of direction goes. But if the time difference is more than 60 to 80 milliseconds (or 60 to 80 feet) we hear a distinct echo. Although we sense a sound as coming from the direction of first arrival, later arrivals add both intensity and a reverberant quality, which is essential for good musical sound.

11 Architectural Acoustics

Between 1895 and 1915, Wallace Clement Sabine, Hollis Professor of Mathematics and Philosophy at Harvard University, laid the foundations of a new science, architectural acoustics. Before Sabine, good acoustical design consisted chiefly of imitating halls in which music sounded good. Poor acoustic design consisted of superstitious practices, such as stringing useless wires across the upper spaces of a church or auditorium.

Architectural acoustics was founded because an opportunity was presented to a remarkable man. The opportunity arose because it was almost impossible to understand speakers in the lecture room of the newly opened Fogg Art Museum. In 1895, the Corporation of Harvard University asked Sabine to remedy this.*

Sabine approached the problems of architectural acoustics with a sharp and inquiring mind, a keen ear, a stopwatch, and an organ pipe with a tank of compressed air as the source of sound. He identified the persistence of sound (i.e., the excessive reverberation) in the Fogg lecture room as the factor that rendered speech unintelligible. He reduced this reverberation by placing felt on particular walls. This, he said, made the room "not excellent, but entirely serviceable."

---

*The story of what he did can be read in his own words, if you can find a copy of his *Collected Papers on Acoustics*, first published by Harvard University Press in 1922, and reprinted by Dover in 1964, but now, unfortunately, out of print.

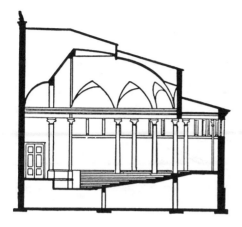

Figure 11-1   Fogg Art Museum lecture hall.

Reverberation Time

Sabine was the first to define *reverberation time*, one important parameter of lecture halls and auditoriums. His definition was the time that it takes, after a sound is stopped, for the reverberant sound level to become barely audible. When accurate electronic measurement of sound level became possible many years later, this turned out to be a fall in sound level of 60 dB. This is how reverberation time is defined today, as we saw in Chapter 10. What reverberation time is considered optimum depends both on the type of music or other sound to be heard and on the size of the room, theater, or other enclosure. Figure 11-2 shows proposed optimum reverberation times for various purposes plotted against room volume.

In considering this table of optimum reverberation times, we should keep in mind that by the time a sound has fallen by 60 dB it is essentially all gone. In a tenth of the reverberation time only a quarter of the sound will be left, and we will be ready to perceive a new sound. Thus, very roughly, the time after which we can perceive a new sound may be around a tenth of the reverberation time of the original sound, and the number of new sounds that can be perceived in one second may be the reciprocal of this number.

With this in mind, consider a conference room with a reverberation time of 0.5 seconds. Only a quarter of the sound will be left after .05 seconds, that is, we will be able to perceive new sounds at a rate of around twenty a second. This is roughly the rate at which new speech sounds can be produced and perceived. Thus, it is reasonable that a 0.5-second reverberation time should be suitable for speech.

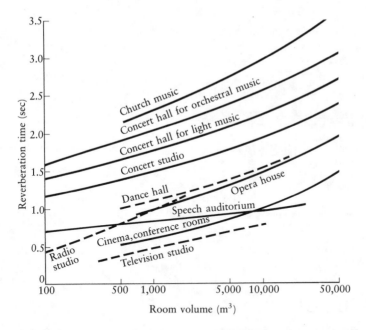

Figure 11-2 The best reverberation time is different for different uses, and for a given use increases with the volume of the enclosure. Church music sounds good in huge, highly reverberant cathedrals. If a room were cubical, the 100-cubic-meter room would be 4.6 meters or 15 feet on a side, the 50,000-cubic-meter room would be 38 meters or 125 feet on a side.

Or consider a reverberation time of 2 seconds for music. New sounds can be perceived at a rate of around five a second, which is consistent with a reasonably rapid succession of notes. But this can be too much reverberation for some sorts of music.

Reverberation has a mellowing effect on sound. Also, it adds to the total sound intensity heard and distributes the sound more uniformly throughout a room. It can even have a confusing effect.

Accurate calculation of reverberation time has been a persistent problem of architectural acoustics. Sabine not only first defined reverberation time, but also devised a useful if not perfectly accurate way of computing it in terms of volume and the fraction of the incident sound that walls and other surfaces reflect.

Sabine's work was arduous. Some was done in an underground room with brick and concrete walls, where, seated in an enclosure, he measured the sound-absorptive properties of window-sized panels of various materials and made other studies. Elsewhere, he spent long nights, waiting for periods quiet enough that he could get an absolute calibration of the sound-absorptive properties of materials by comparing their effect with that of an open window. Sabine measured the sound absorption of a host of materials. He also studied sound conduction and means for isolating practice rooms acoustically, a persistent problem in conservatories.

Mostly, Sabine was called on to cure or ameliorate the bad acoustics of halls built by presumably respectable architects who were either grossly ignorant of the needs of listeners or who simply didn't care about them, traits still all too common among architects today. But Sabine was also able to do an original acoustical design for Symphony Hall in Boston, built in 1900 to replace the old Music Hall. Today, Symphony Hall is one of the very few outstanding concert halls in the world.

Problems of Architectural Acoustics

Research in architectural acoustics waned in the land of its birth, but excellent work was instituted in Germany and other countries. There are two general problems in architectural acoustics, and each has many aspects. One problem is, What do we want? What enables performers to play well? When they do play well, what is it that makes them sound good? The other problem is, How can we attain both what is good for the performers and what is good for the audience?

Figure 11-3 Symphony Hall, Boston.

Noise is important in concert halls, as noted in Chapter 9. Concert-hall design requires great attention both to excluding external noise and to not producing noise. For example, air-conditioning systems are often excessively noisy in offices, and sometimes are in concert halls.

The performer needs to hear reflected sound, as noted in Chapter 10. This has been the subject of several recent studies, both of performers' preferences for various halls and of performance under laboratory conditions. Although any amount of reverberant sound may satisfy soloists, satisfactory ensemble playing depends on early reflections of sound from behind and above the performers. Each player must hear all the rest of the players by means of reflected sound that is not too much delayed. Good transmission of sound from the orchestra to the audience will be of little avail if the players can't play well and comfortably together. It is common lore that players and even audiences like to "feel" music through a wooden (as opposed to a concrete) floor.

The problem of out-of-doors performances, such as those in the Hollywood Bowl, are as old as the Greek theater. Such performances are plagued by noise. Although it is easy to provide reflecting surfaces near the orchestra to enable the performers to hear one another, it is impossible to provide reverberant sound to the audience, at least, not without amplification and artificial reverberation. Without electronic aid, outdoor music *can't* sound as good as music in a concert hall. It may be fun, but the reasons it is fun aren't good acoustics.

In this chapter we will consider in some detail both the physical aspect of measuring and predicting the transmission and decay of sound in concert halls, and the psychological aspect: What makes a good hall good? Sabine appreciated both of these aspects. He experimented to find the preferred reverberation time for musical performance, and he knew that the optimum reverberation time for music is longer than that for speech. Sabine also understood the objectionable quality of echoes and knew how to prevent or cure them.

A Case History: Philharmonic Hall, Lincoln Center

Much progress has been made since Sabine's time. Some has hinged on new psychoacoustic understanding (e.g., of the precedence, or Haas, effect mentioned in Chapter 10). Electronic means for generating and measuring sound have been most useful. The electronic digital computer has been of great help. But these resources would have meant nothing without the sharp, inquiring minds of a few men and women who have, through the years, advanced the science of architectural acoustics.

How far has it advanced? An example shows that assertions can be misleading. Philharmonic Hall in Lincoln Center, Manhattan, opened on

Figure 11-4   Philharmonic Hall, Lincoln Center, as designed by Leo Beranek.

September 12, 1962. It was a disaster. Yet, in July of 1962, Leo L. Beranek, of the architectural firm of Bolt, Beranek, and Newman, wrote in the preface of his book *Music, Acoustics, and Architectural Design,* "The climax of this volume is the description of the care taken in planning the Philharmonic Hall in Lincoln Center. Lady Luck has finally been supplanted by careful analysis and the painstaking application of new but firmly grounded acoustic principles."

What went wrong in the original design of Philharmonic Hall, and why?

The *why* appears to be that great attention was given to matters that Beranek deemed to be of great importance, whereas little was given to other matters that also proved to be of great importance.*

---

*Beranek was so convinced of the correctness of his theories that he did not even bother to build and test a model of the hall. Models have been made since Sabine's day, and his book shows pictures of sound propagating through a model auditorium. What Sabine could do crudely can now be done well. The wavelength of sound in a model should preserve the same relation to dimensions as in the actual hall; for example, in testing a tenth-scale model, a frequency of 500 Hz should be represented by a frequency of 5,000 Hz.

The *what* includes almost everything. (The details can be found in papers by Manfred Schroeder and his colleagues.) There were echoes at some seat locations. The members of the orchestra couldn't hear themselves and others play, because the wall behind them was absorptive. There was a lack of subjectively felt reverberation. There was inadequate diffusion of sound through the hall. Worst of all, there was an apparent absence of low frequencies: it was difficult to hear the celli and double basses.

This was not apparent in measurements of the total sound energy that reached the hearer during the period of one second after the first arrival of direct sound. Figure 11-5 shows the total energy that reached the hearer in various octave bands of frequency. Though the energy has fallen somewhat in the octave from 125 to 250 Hz, the drop is only about 5 dB, which isn't much.

There *was* indeed an initial lack of low-frequency sound, both in the sound that arrived directly from the stage and in sound reflected from a suspended ceiling made up of a large number of adjustable panels, or "clouds."

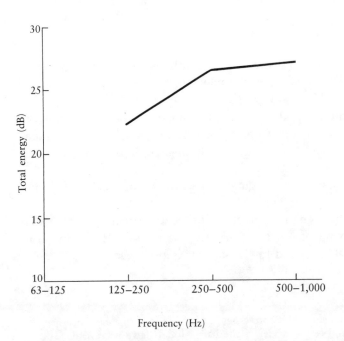

Figure 11-5   Total energy from the stage of Philharmonic Hall in decibels, averaged over five main floor positions, for three octave bands. For the band from 125 to 250 Hz, the received energy is about 5 dB less than that for the band from 500 to 1,000 Hz.

Figure 11-6 shows the deficiency in the direct sound from the stage. The relative energy of the direct sound in dB is plotted against frequency for distances of 7, 13, and 31 meters from the stage, along the center aisle of the main floor. There is a pronounced dip between 100 and 200 Hz, and it becomes greater at greater distances from the stage. The original floor of the hall wasn't raked much; that is, it didn't go downhill very much toward the stage. The sound skimmed along above the rows of seats. Because the space between rows of seats tended to act as a resonator, at low frequencies the sound waves were bent upward, away from the audience. The way that the sound waves were reflected from the regular, successive rows of seat backs (diffraction) added to the loss of low frequencies.

As for sound reflected from the ceiling, Figure 11-7 shows the average energy reflected from the "clouds" during a short interval centered on the arrival time from the clouds. For the octave from 125 to 250 Hz, the average energy has fallen about 11 dB below its value at the octave from 500 to 1,000 Hz.

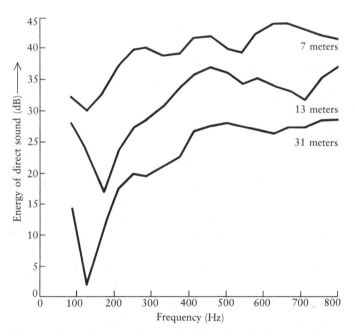

Figure 11-6 Energy of sound going directly from the stage of Philharmonic Hall to various locations along the center aisle on the main floor, as a function of frequency. At 31 meters from the stage, the sound at 130 Hz is about 25 dB below that at 500 Hz.

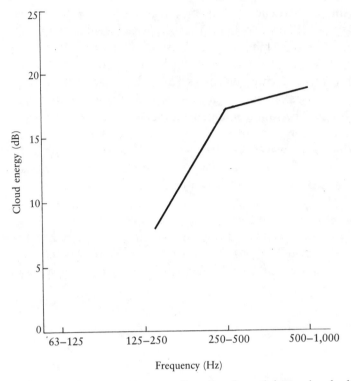

Figure 11-7 Energy in octave bands reflected from clouds, like those in Philharmonic Hall, averaged over five main floor positions. This measurement was made in a model. The energy reflected in the band from 125 to 250 Hz is about 10 dB below that reflected in the band from 500 to 1,000 Hz.

Why was so little low-frequency energy reflected from the clouds? Because they weren't large enough! A flat surface acts as an effective reflector only if it is large measured in wavelengths of sound. The clouds weren't large enough to reflect effectively sounds having frequencies below 300 Hz.

The curve of Figure 11-6 shows how little direct sound of low frequency reached the main floor. Figure 11-7 shows how little was reflected from the clouds. Why, then, does Figure 11-5 show a considerable amount of low-frequency sound energy finally reaching the listener? It reached the listener only after repeated reflections, some above the clouds, which low-frequency sounds passed through, and some after traversing the hall many times. This sound arrived so late that the listener failed to associate it with the notes that the celli and basses were playing. It became, in effect, a background noise, detached from its musical source.

There were many early efforts to patch up Philharmonic Hall. A solid stage enclosure was built so that the orchestra could hear itself play. The clouds were realigned to form an essentially continuous ceiling. Scattering elements were put on the side walls to give better sound diffusion. New, less absorptive seats were installed on the main floor. The front of the balcony was tilted, and absorptive material was placed on the back wall to diminish echoes. This made the hall somewhat better, though the reverberation time became rather low (about 1.85 seconds).

Finally, Philharmonic Hall was completely redesigned by one of the few first-rate American experts in architectural acoustics, Cyril Harris of Columbia University, and became Avery Fisher Hall.

The story of Wallace Sabine is that of a triumph of new scientific understanding. The story of Philharmonic Hall is that of an expensive disaster based on incomplete knowledge. However, there is a happier side to architectural acoustics, well illustrated by the work of Manfred Schroeder, whose contributions illustrate what progress in architectural acoustics can be.

Figure 11-8   Avery Fisher Hall, Lincoln Center, on opening night.

Recent Research in Architectural Acoustics

A persistent problem in designing concert halls is the accurate prediction of reverberation time. Sabine gave a simple formula for the reverberation time T, measured in seconds, as

$$T = \frac{13.8L}{va}$$

Here L is the mean free path between successive reflections of sound waves, v is the velocity of sound, and a is the sound-absorption coefficient, which is zero for perfect reflection and unity for complete absorption.

Sabine assumed that the mean free path L was proportional to the cube root of the volume. However, it was already known from the kinetic theory of gases that under what is called an *ergodic* condition, in which the sound traverses all possible paths, the mean free path is given by

$$L = \frac{4V}{S}$$

Here V is volume, and S is the internal surface area of the volume.

A problem with Sabine's formula is that it predicts a finite reverberation time for complete absorption ($a = 1$). In 1929, K. Schuster and E. Waetzmann, and in 1930, Carl F. Eyring remedied this by proposing a revised formula of

$$T = -13.8(L/v)/\ln (1 - a)$$

The Sabine/Eyring formulas were accepted for more than half a century, but in the 1960s, accurate electronic measures of various halls, new and old, cast doubt on them, and on other approximations as well.

One approach to predicting reverberation time is to use a digital computer to trace many sound paths, or rays, that travel out in different directions from a sound source. Each ray is successively reflected from various points on the walls of an enclosure. A ray will sometimes be reflected by absorptive material and lose energy; sometimes it will be reflected without appreciable loss of energy by such materials as wood or plaster.

When ray tracing was first tried, computing was slow and expensive. Thus, in 1970 Schroeder traced rays in irregular two-dimensional structure in which one of four linear boundaries absorbed some fraction of the incident sound. Figure 11-9 shows the enclosure and the way the sound actually decays in it, with decay curves according to Sabine's and Eyring's formulas for comparison. In using these formulas the value of the absorption coefficient a was taken as the value averaged over all boundaries. Ray tracing gave a reverberation time that was 0.6 of that predicted by Sabine's formula, and about 0.7 of that predicted by Eyring's.

In various papers published from 1982 through 1988, E. N. Gilbert of Bell Laboratories gave exact solutions for several simple room shapes and compared these with various approximations. For reasonable reverberation times, deviations from those predicted by the approximate formulae were smaller than deviations Schroeder had found in his two-dimensional examples.

Today ray tracing software can be run on desktop computers. Detailed data on multiply reflected sound rays can be obtained. For a loud click at one spot in the room, it is possible to compute how much sound will arrive at a given point in a given solid angle from a given direction in a given short time interval at a given time after the initial sound — that is,

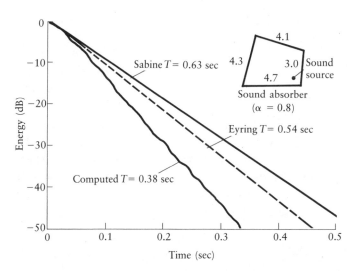

Figure 11-9   Decay of sound computed by ray tracing for a trapezoidal enclosure with absorbing material on one wall. This computed curve corresponds to a reverberation time (α) of 0.38 seconds. The reverberation time calculated according to Sabine is 0.63 seconds; that according to Eyring is 0.56 seconds. The Sabine and Eyring calculations do not take into account the fact that all sound absorption is on one wall.

after multiple reflections, where the sound will be coming from as a function of time.

Reverberation time is important in measuring the fraction of incident sound reflected from various materials. The material to be tested can form one wall of an enclosure in which the other walls reflect sound without appreciable loss. If the reverberation time can be computed accurately as a function of the coefficient of absorption, that coefficient can be computed from the measured reverberation time.

Important as the evaluation of acoustical materials is, we are more interested in the relation of reverberation to the acoustical quality of a concert hall. How good does an orchestra sound in a particular hall? Judgment is difficult, for a particular orchestra will play in different halls at different times. Further, the orchestra may not play in just the same manner in halls with different acoustical properties. Is there any way (outrageous thought!) to bring different halls to one listener at the same time? There is!

In 1967, Manfred Schroeder and Bishnu Atal showed how two loud-speakers could be used to produce an apparent sound source that lay to the left or right of both speakers. In 1969, P. Damaske and V. Mellert showed how to make use of this effect in producing a "perfect" stereo system. Sounds were picked up from the two ear canals of a dummy head, complete with pinnae. Figure 11-10 shows such a dummy head. These signals were then filtered and mixed properly, and fed to two loudspeakers in an anechoic room. Figure 11-11 indicates how the sound was reproduced from the two-track recording. The sounds from the two speakers could re-create in the ear canals of a dummy head exactly the same sound pressures that had been recorded when the head was exposed to live sound during the recording.

Schroeder and Atal's work had important applications for architectural acoustics. Suppose that you made a two-channel recording of the sounds in the ear canals of a dummy head "seated" in a concert hall. From the recording you could re-create for a listener in an anechoic room *exactly what the dummy head "heard" in the concert hall*. By switching from a recording in one hall to a recording in another hall, you could compare the two.

Schroeder and two collaborators, D. Gottlob and K. F. Siebrasse, undertook to compare more than twenty European concert halls. They managed to obtain a multichannel tape recording of Mozart's *Jupiter* Symphony played by the BBC orchestra in an anechoic room. This tape they played back over several loudspeakers on the stages of various concert halls. In each hall they made two-channel tapes of what the dummy head heard when seated in several locations. From these recordings they re-

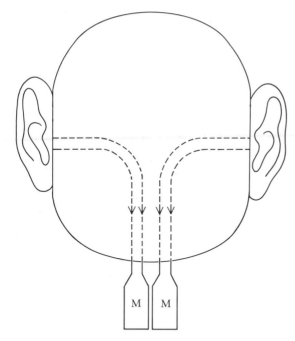

Figure 11-10 A dummy head with pinnae. Microphones (M) pick up the sound pressure in the ear canals. Two-track stereo recordings can be made in a concert hall by "seating" the dummy head in the hall.

created for a listener seated in an anechoic room exactly what the dummy head "heard" in various concert halls. Schroeder comments,

> I will never forget the moment when, comfortably seated in the Goetingen "free space" room, I first switched myself from Vienna's famed Musikvereinsaal to Berlin's Philharmonic. The acoustical differences of these halls, always believed to "exist," stood out in a manner so vivid that it is difficult to put into words.

By analyzing the judgments that various listeners made "listening" in many positions in many halls, Schroeder and his collaborators learned what listeners like:

1. They liked long reverberation times (below 2.2 seconds).
2. They liked the sounds to differ at their two ears. The more nearly alike (correlated) the sounds were at the two ears, the less they liked them.
3. They liked narrow halls better than wide halls. Perhaps this is another expression of a preference for different sounds at the two ears. In a

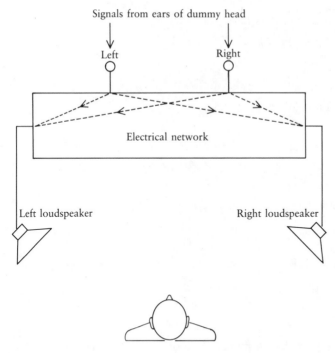

Figure 11-11 If the two-track recording made with the dummy head shown in Figure 11-12 is played back by means of two loudspeakers in an anechoic room, with the use of a proper network so that some signal from each soundtrack reaches each loudspeaker, the pressures in the ear canals of a real (or dummy) head will be exactly and only those recorded in the ear canals of the dummy head seated in the concert hall. The fraction of the signal fed from the microphone in each ear of the dummy head to the left and right loudspeakers must change properly with frequency.

wide hall the first reflected sound rays reach the listener from the ceiling. In narrow halls the first reflections reach the listener from the left and right walls, and these two reflections are different.

In his doctoral work, completed at Stanford in 1984, Jeffrey Borish further added to our knowledge. He found that we are pleased to hear early reflected sound arrive strongly from somewhat to the left front and somewhat to the right front, as it does in a long, narrow, rectangular concert hall with a high ceiling. Through a "method of images"—an alternative to ray tracing—Borish used a computer to calculate the arrival directions and times of once- and severally reflected sounds for a number of auditorium shapes. Fan-shaped auditoriums wider at the back than at

Figure 11-12   Dummy head with D. Gottlob (left) and K. F. Siebrasse (right).

the stage are objectionable, because reflections from the walls miss the audience entirely. Low ceilings are also objectionable because the earliest reflected sound comes from above rather than from the sides.

Roughly, then, this is what a good hall must do: It must have a long enough reverberation time (below 2.2 seconds). Early reflected sound should reach us from left forward and from right forward, differently delayed. Multiple reflections, necessary for adequate reverberation time, must mix the sound thoroughly and diffuse it throughout the hall. And these criteria must be approximated, at least, for most seats.

Much mathematics has been expended in designing textured walls or ceilings with bumps and wells that will scatter incident sound in many directions rather than reflecting it as a mirror does. Figure 11-13 shows a structure designed by Schroeder in which the depths of long, narrow wells in a ceiling are based on "quadratic residues" derived from number theory.

One problem we have not mentioned: The bigger the hall, the harder it is to make it acoustically good. Yet a large seating capacity somewhat offsets the ever-growing costs of performance. Isn't there some new, untried magic that will give us wonderful acoustics in a stupendous hall? Donors and the architects they employ are eternally gullible. The acoustics of Louise M. Davies Hall in San Francisco were initially unacceptable and have been only partially remedied through reflectors hung from the ceiling. Will a new hall soon to be built in Los Angeles be better?

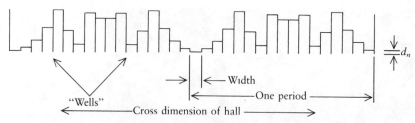

Figure 11-13 In a structure designed by Manfred Schroeder, a sequence of "wells" scatters incident sound widely without loss, rather than reflecting it like a mirror. The scattering changes somewhat with frequency but is successful over a wide band of frequencies. Material with a similar surface may be used in concert halls of the future to insure diffusion of sound throughout the hall and to distribute the sound so that the sounds reaching the two ears are different.

Actually, there is a type of acoustical magic: the augmentation of the sound wave through various electronic means. This may involve augmented reflection of sound from walls, as in the Royal Festival Hall in London. It may involve picked up and delayed (reverberant) sound emitted from loudspeakers on the periphery of the hall. It could even involve the augmentation of the direct orchestral sound itself, as in a public address system. In the United States designers of large and expensive halls shy away from electronics, perhaps rightly so. A failure of electronically augmented sound would be castigated far more severely than a fiasco not involving electronics.

12 Sound Reproduction

The electronic reproduction of music has become more common than live music. Popular music accounts for the vast majority of recording sales, but recordings of classical music are extensive. Once such music could be heard only in the excitement (or boredom) of a concert hall. Today a symphony orchestra that does not make money from recordings, TV, or films must survive on charity or government subsidies. Yet, people hear more music, and a greater variety of music, than even before.

The problem of adequate quality in recorded music is twofold. There is the problem of making a faithful two-track (stereo) recording. This has been solved by recording sound digitally and by the advent of the compact disc. The signal-to-noise ratio can be better than 90 dB. The frequency response extends from 20 to 20,000 Hz.

Ideally, a signal of bandwidth B can be represented *exactly* by 2B binary numbers a second, successive numbers representing successive *samples* of the signal wave. A sample is simply the amplitude of the signal wave at any given moment. Ideally, then, a signal with frequencies up to 20,000 Hz could be represented by 40,000 numbers a second. Actually, in compact discs 44,100 numbers a second are used. Each successive number, or sample, is represented with adequate accuracy by sixteen binary digits. These are encoded ingeniously as sequences of pits along a 3-mile spiral path on the compact disc. The recording is read by means of a laser beam whose diameter is about 2 millionths of a meter (less than a ten-thousandth of an inch).

The other problem is that of reproducing recorded sound with satisfactory accuracy. This is easiest with headphones. We now have Walkman-like portable disc players that give excellent frequency response and amplitude range. But even this isn't the same as listening to a live performance in a concert hall.

Sound quality isn't everything. We can enjoy conventional works that we have heard before even when they are poorly reproduced—as long as they are well played. But music, and especially computer-generated music, may depend on deep bass tones, or stereo effects that can take sounds on a ghostly course around the room. Such sounds call for a good stereo system that will produce them well.

Reproducing the sound of a large orchestra as loudly and clearly as we wish depends on what we ask of sound and where we hear it. On April 27, 1933, Harvey Fletcher and his co-workers demonstrated a three-channel system that carried the sound of the Philadelphia Orchestra, conducted by Alexander Smallens in the Academy of Music in Philadelphia, to three loudspeakers on the stage of Constitution Hall in Washington, D.C. During the demonstration Leopold Stokowski manipulated the electronic controls from a director's box in the rear of Constitution Hall. Figure 12-1 shows the placement of the microphones in the Academy of Music and of the loudspeakers and the director's box in Constitution Hall.

The performance of this system was outstanding by any standard. The bandwidth was from 40 to 15,000 Hz. The range between noise in the absence of signal and the highest average signal power was 75 dB (70 million times). The largest undistorted sound-output power that each of the three loudspeakers could deliver was 135 watts, or 405 watts for all three loudspeakers. Each loudspeaker (see Figure 12-3) consisted of a folded horn woofer with a square aperture about five feet on a side to handle frequencies from 40 to about 300 Hertz. This was surmounted by a cellular horn about two feet high and three feet wide to handle higher frequencies.

The acoustic power that this system could deliver was several times that which a large orchestra can produce, which is about 70 watts. The story is that Stokowski kept turning the sound up louder and louder, and that the controls had to be modified to keep him from overloading the

Figure 12-1 (*Opposite page*) On April 27, 1933, Fletcher and his co-workers transmitted the sound of a symphony orchestra playing in the Academy of Music in Philadelphia (*top*) to Constitution Hall in Washington, D.C. (*bottom*). The sound was picked up by three microphones between orchestra and audience, and was reproduced by means of three loudspeakers on the stage of Constitution Hall. In a director's box near the back of Constitution Hall, Leopold Stokowski adjusted sound levels during the performance.

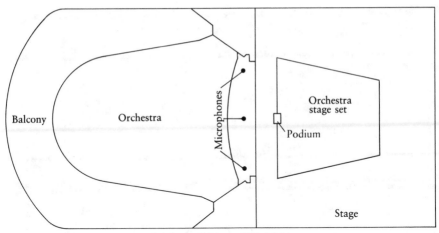

Academy of Music, Philadelphia

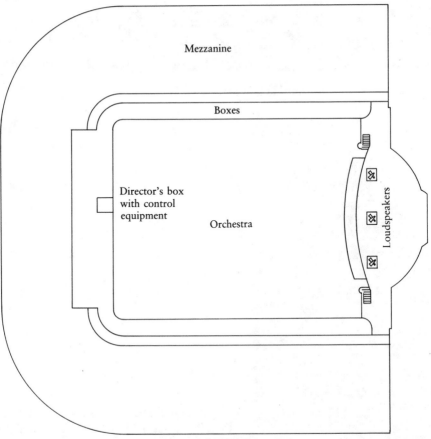

Constitution Hall, Washington, D.C.

Figure 12-2 Constitution Hall, Washington, D.C.

Figure 12-3 The type of speakers used in the 1933 transmission from Philadelphia to Washington.

system and blasting the audience. Whatever the truth of this may be, by all accounts the effect was good. The sense of the sources of the sounds of individual instruments cannot have been preserved, but in most seats in a large hall we are so far from the stage that we scarcely sense more than left, center, or right. Mostly we are enveloped in a field of reverberant sound from the walls and ceiling, and it seems that little was lost because the sound came from three loudspeakers placed appropriately on the stage.

In experiments described in Chapter 11, Manfred Schroeder obtained a very convincing sense of the presence of an orchestra by playing a multichannel recording through loudspeakers placed on the stages of concert halls. Schroeder's reproduced sound should have been better than Fletcher's, for the recording was made in an anechoic room, whereas in Fletcher's system some reverberation from the first hall must have gotten into the sound sent to the second.

Problems of Home Stereo Systems

When we think of sound reproduction, we usually think of reproducing the sound of a singer, instrumentalist, ensemble, or orchestra in our own home. This can be done — to a degree. In Chapter 11, I described a perfect two-track stereo system, in which sound pressure recorded from in the "ear canals" of a dummy head is faithfully reproduced in the ears of a listener seated in an anechoic room. We should note three limitations.

First, this system can work perfectly *only* in an anechoic room. In any other enclosure, the reverberation of the enclosure is added to that of the hall in which the sound was recorded. This may not be very important, because the precedence effect insures that you will at least hear the sound coming from the right direction.

Second, you can't turn your head very much.

Third, you must be equidistant from the two loudspeakers, and at just the right distance from them. If you move close enough to one loudspeaker, any sound that comes from both speakers will seem to come from the speaker to which you are nearest. If you move a little, you may hear sound from both speakers, but phantom auditory images of compact sound sources will move and become blurred.

Despite these limitations of the "perfect" two-channel system, something may be gained by a crude use of one of its features, that of driving each speaker from a mixture of the two recorded sound channels, as shown in Figure 11-11. Stereo manufacturers supply networks that do something of the sort, and the effect seems to be an increase in "presence," or being surrounded by sound. We can even feel surrounded by sound from a monaural source, as indicated in Figure 12-4. Of course, this cannot

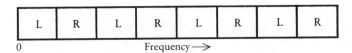

Figure 12-4 We can get a pseudo-stereo signal by feeding alternate frequency bands of a monaural signal to left and right loudspeakers or to left and right headphones. More striking results can be obtained by dividing the monaural source in more complex ways—for example, by using artificial reverberation. Such a system will spread the apparent sound source of an individual instrument out through space.

approximate the "perfect stereo" effect, for the two sound channels that are mixed were not recorded with the ears of a dummy head.

Ludwig II, king of Bavaria, had an opera house with one seat for an audience of one—himself. Many of us would prefer to share performances, live or recorded, with others. Is it possible to share electronically reproduced sound "equally"? With headphones, perhaps; we will come to that later. Using loudspeakers, it is possible in principle, but not in practice, for the following reasons.

Imagine a room-sized enclosure built in a concert hall, with musicians playing on the stage outside the enclosure, and yourself seated inside the enclosure.

If all walls, the floor, and the ceiling are solid and soundproof, without apertures, you will hear nothing.

If we cut one hole in a wall, you will hear all the music as coming from that hole, as you would from a single loudspeaker.

If we cut two holes in the wall, you will get an effect much like that of a conventional stereo system. The music will sound like good stereo if you sit equidistant from both holes, but you will never hear anything coming from a direction to the right or the left of the two holes (loudspeakers), as you can with the "perfect" stereo system in which sound is picked up by a dummy head, and you will have no sense of sound coming from above or below the holes.

If we cut a hole in every wall, you can sense sound as coming from any direction, but again, if you move close to one hole, you will hear sound as coming from that hole, and unless you are equidistant from the four holes, you will not hear sounds as coming from their original compact sources in the correct directions.

Only if we cut many holes in the walls and ceiling can we hear sound within the enclosure pretty much as we would hear it in the absence of the enclosure. We could approximate this effect by putting many, many loudspeakers on the walls and ceiling of a room, and feeding an amplified signal to each loudspeaker from a microphone placed in an analogous position in a concert hall. Such a multichannel system would be costly, unwieldy, and impractical.

Recall from Chapter 2 that a sound wave consists of both motion of the air and compression or expansion of the air. At any point, a sound wave, no matter how complicated, disturbs the air in just four ways: by a fluctuating pressure, by a fluctuating up-and-down velocity, by a fluctuating forward-and-back velocity, and by a fluctuating left-and-right velocity. All of these disturbances can be measured by using a microphone sensitive to pressure together with three microphones sensitive to velocity in one direction.

Suppose we use four loudspeakers, which can be at the vertices of a three-sided pyramid (tetrahedron), and drive each from a particular linear combination of the signals from the four microphones (see Figure 12-5). The speakers can produce, at one point only (which can be at the center of the tetrahedron) the exact fluctuations in pressure and velocity that the four microphones responded to.

Thus, at one point, surrounded by the loudspeakers, the fluctuations in air pressure and velocity will be exactly the same as those at the microphones in the recording studio. If you place your head at this point, and your head isn't too large, you would hear exactly what you would hear at a particular point in the studio.

Techniques of Sound Recording

We should not confuse the four-speaker system described above with what is called quadraphonic sound. Quadraphonic sound was intended to improve the sense of direction of sound from in front of the listener, and to

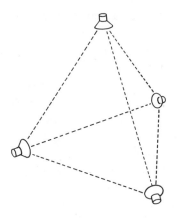

Figure 12-5 The exact sound velocities and pressure at a point in a recording studio can be reproduced at a single point in an anechoic room by means of four loudspeakers placed at the corners of a tetrahedron. Four microphones in the studio pick up signals proportional to (1) the up-and-down velocity of the air, (2) the forward-and-back velocity, (3) the left-and-right velocity, and (4) the pressure. The signal that drives each loudspeaker is a particular linear combination of these four signals.

give a sense of sound from behind the listener, too — of being surrounded by sound, direct or reflected.

Quadraphonic sound intended for home use employed four speakers. It did not necessarily call for four recorded channels. In matrix quad, signals from four microphones or constellations of microphones were combined artfully and recorded as two independent channels. The four channels for driving four (or more) speakers were then derived from the two recorded channels. Matrix quad works (sort of) because the signals recorded from four microphones aren't completely independent.

Quadraphonic systems didn't survive in the home audio market, perhaps because of their limitations and the added cost. However, multiple-speaker systems have taken over in movie theaters.

Seventy-millimeter films have six sound tracks and six or more speakers. Thirty-five-millimeter films have two sound channels, derived, as in matrix quad, from four microphone channels, according to a patented (and standardized) process, Dolby stereo. From the two recorded tracks, four speaker signals are derived, for speakers behind the screen and left, center, and right, and for "surround" speakers toward the sides and rear of the theatre.

Dolby stereo is available for the home. You can now buy or rent movies on half-inch VHS-format videocassettes that have two audio tracks (in analog) that are encoded in Dolby stereo. You can purchase add-on decoders to connect your videocassette player to your home stereo system. These will decode the two channels into the left-center-right-surround convention used in movie theaters. Many stereo systems on the market now have Dolby stereo decoding built in from the beginning. Arts' Home-THX system provides a layer on top of Dolby stereo that equalizes home speakers and attains something of the effect of sound in a movie theater.

In recording it is common to derive two signals from a monaural channel (recording a solo instrument) so that in stereo reproduction the sound will seem to come from a particular place. Q-Sound provides such a technique. Roland provides a box, RSS (Roland Surround Sound), that will convert a monaural channel to signals for two or four speakers.

Both the Roland and Q-Sound systems make extensive use of the simulation of perceptual cues for spatial positioning. They attempt to simulate and, in some cases, to exaggerate the natural cues used by the ear to determine the locations of sound. Since we cannot re-create exactly the three-dimensional sound field of an acoustic instrument being played in our living room, some compromises are necessary. These systems can be made to work quite well when heard through headphones, as long as the listener does not move his or her head. When the systems are presented free-field by loudspeakers, there is often a "sweet spot" at a certain position in the room where the effect works quite well, but in other spots the effect does not work as well.

Beyond this, proprietary networks in some home sound systems will drive two stereo speakers in such a way that the sound seems to come from many directions.

We should note that one important purpose of recording and sound systems is to sell recordings. Recordings will sell if they please listeners. The techniques used in recording popular music have been developed with this in mind. Popular music is recorded on many tracks, usually 24 for rock music, and as many as 32 are now commercially available. Up to four 24-track tape machines have been slaved together to produce 96-track recordings. In major motion picture production, use of over one hundred tracks is common. In recording small groups, a different microphone is put near each instrument, or a signal can come directly from the instrument itself, as from an electric guitar or a digital keyboard. A voice track, or other solo track, may be recorded separately by a performer who listens to the rest of the recording.

When the various tracks are finally assembled, the resulting combination of sounds is a new, complex sound that never before existed, that could not have been heard before. The listener could not hear that sound in a live performance. Each instrument and vocalist has been recorded very close up, and it is impossible to place the listener's ear at any point in space where he or she could hear the combination of sounds provided by the microphones. Indeed, some of the performers are recorded at different times or even in different places. Thus, multitrack recording results in new syntheses of sound, and not simply in the reproduction of a performance. What we are used to hearing in popular music recordings is an unnatural combination of separate, discrete sounds placed in an artificial stereo environment.

This is fundamentally different from our idea of a classical music recording that strives to capture faithfully the sound of an actual performance. Well, few do. Rather, they strive to give the *impression* of an actual performance. Often close miking is used to enhance the "presence" of a soloist for a certain passage. Also, classical music is the *most highly edited* music on the market today, more so than pop music. Except for "live" recordings, the performances you get on modern classical recordings have had every single bad note edited out and replaced by a picture-perfect note from some other take or performance, so that the result is inhumanly perfect. That is what the consumer has demanded of the recording industry, and the industry has responded. Classical recording studios use hardware such as SonicSystem™ to help them speed up this elaborate editing process.

What people want changes with time. I can remember when the bass boom of the Majestic radio or jukebox was a big thing in music. Many people now enjoy feeling immersed in sound, no matter where it is coming from. Artful stereo can do this.

Artificial reverberation is important in adding a natural quality to sounds. The reverberant sound can be different in the two-stereo or pseudo-stereo channels. In the past, reverberant sound was sometimes produced by driving a large brass plate with a loudspeaker mechanism, picking up signals from various portions of the plate, and adding these to the original sound. Today, reverberation of adjustable duration and quality is easily (and inexpensively) produced by commercial digital reverberation devices. If we drive several loudspeakers with several different reverberant signals we get a sense of being surrounded by sound, much as in a concert hall, where the sound reaches our ears by reflection from different directions by different walls.

Indeed, reverberators and loudspeakers can be used to add suitable reverberant sound to halls of moderate size that have inadequate reverberation. Microphones on the stage area pick up sound, which is suitably delayed and amplified and fed to speakers on the periphery of the hall. This gives good results only if the hall is of reasonable shape, and if there is adequate direct sound. In one way such additional electronic reverberation is better than actual built-in reverberation. The reverberation can be increased to give a rich effect in listening to music and reduced to allow the clear perception of speech.

Figure 12-6 A jukebox.

Reverberant sound can give a sense of immersion in sound in a recording played in the home, with only one independent sound channel, or better, with two. What we can't get is an accurate sense of discrete sound sources. An orchestra, a piano, or a singer seem to envelop us. This is what some purists object to.

There *is* one way to "hear it like it is": with two channels and headphones.

With headphones we easily sense the difference between single-channel (monophonic) sound and stereo, but ordinary stereo sound seems to be inside our head. We don't externalize such sounds, because they aren't recorded in a way that will allow this. But if the two channels recorded from the ear canals of a dummy head are fed to good headphones through proper filters (to compensate for the fact that the pinna gets into the process twice), the two-channel sound can be heard as external. Tone controls can be used to approximate the proper filters.

Sometimes, dummy-head recordings lead to confusion between sounds in front of us and sounds behind us. Recording from tiny microphones in our own ear canals is better. We hear just what we would in a concert hall, as long as we don't turn our heads. If, in our home, we *really* want to share the same properly reproduced music among a group of listeners, we must ask them to wear headphones, and all to stare fixedly in one direction. But will *you* hear properly what has been recorded using *my* ears? Alas, we can't try this with stereo as it is now recorded.

Nonlinear Distortion

Stereo enthusiasts often claim that their golden ears diagnose faults that measurements do not disclose. They are much concerned, and rightly, with nonlinear distortion, and some, rightly, prefer old vacuum-tube power amplifiers to badly made but elaborate transistor amplifiers. This is because the good vacuum-tube amplifiers and the bad transistor amplifiers have different *kinds* of nonlinear distortion.

"Good" and "bad" nonlinear distortion are illustrated in Figure 12-7. In the "good" distortion (part A), the actual curve of output voltage plotted against input voltage departs from a straight line smoothly and gradually. In the "bad" distortion (part B), the curve of output voltage plotted against input voltage bends sharply.

The "good" distortion may actually improve the sound of a solo instrument, making it richer. That is why such nonlinear distortion is sometimes used in processing the sound of a single instrument in recording studios, before the sounds of different instruments are mixed together. Even "good" distortion is bad when the waveform distorted is that of two or more instruments playing different notes, but it may not be very bad.

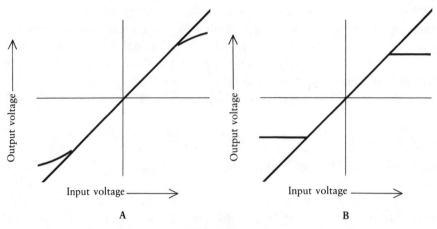

Figure 12-7 "Good" nonlinear distortion (**A**), contrasted with "bad" nonlinear distortion (**B**). In each case the curve of output voltage plotted against input voltage departs from the ideal straight line. But in "good" distortion the departure is smooth and gradual; in "bad" distortion the departure is sudden, putting a sharp bend in the output waveform, adding high frequencies and giving a distorted sound.

The "bad" distortion is bad for everything. Figure 12-8 shows what it does to a sine wave. It cuts off the smooth peaks and valleys and leaves a squarish wave that sounds harsh and distorted to any ears, golden or not.

There is good reason to prefer vacuum-tube power amplifiers with "good" distortion to transistor amplifiers with "bad" distortion. Further, it is easier to get very "bad" distortion with transistors than with vacuum

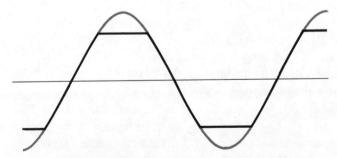

Figure 12-8 The effect of "bad" nonlinear distortion on a single sine wave. The "bad" distortion cuts off the peaks and valleys of the sine curve, leaving a squarish wave that sounds harsh and distorted.

tubes—through bad design. Some senseless enthusiasts go so far as to resurrect vacuum-tube preamplifiers (which amplify the very low-level signal from a microphone) and use them to replace *good* transistor preamplifiers. I believe they do this because they have been misled by the stereo trade, which makes and quotes almost meaningless measurements that disagree with what our ears tell us. But the remedy here should be good measurements, not a refusal to measure at all.

As we noted in the discussion of masking (see Chapter 9), false signal components that lie close in frequency to strong signal components will be masked, and we won't hear them. However, even a little distortion in a frequency range where there is no signal stands out like a sore thumb. Distortion is commonly measured by using only one or two sinusoidal tones as an input. It would be far more realistic (though more difficult) to import one of the techniques used for measuring distortion in the broadband telephone amplifiers used in "carrier" systems to amplify simultaneously many telephone channels. One technique uses as an input signal a noise signal that contains all the frequencies that the amplifier must amplify *except* in one narrow range of frequencies. With such an input we can measure in the output of the amplifier the distortion power in the range of frequencies that aren't present in the input. This technique has indeed been tried in recent years, but it is not in common use.

In the early days of high-fidelity reproduction, the infant art was regarded as at the forefront of engineering. Pioneers such as Harvey Fletcher and his colleagues at Bell Laboratories were highly respected in the general community of scientists and engineers, as was Harry F. Olson at RCA. Their work was also appreciated and followed in universities.

Today, advances in sound quality (and in sound synthesis) lie chiefly in the hands of brilliant researchers in enterprises of medium and small size. Though good work is done in computer music centers on various university campuses, music and sound aren't high priorities in the academic world nor in the funding patterns of either private foundations or government sources.

This is less the case in some European countries and in Japan. There is fine acoustical work being done in various foreign universities and institutes. Boulez's IRCAM is financed by the French government, and electronic music and computer music centers in other European countries also have government support. Even though ill funded, computer music centers have been founded at various American universities.

The digital revolution, which began in computers and communication, has brought many talented musicians, engineers, physicists, and psychologists to the study of sound and sound quality. A renaissance is under way, whether or not those who might fund it are aware of what is happening.

13 Musical Instruments, Analysis, and Synthesis

A musical revolution has occurred over the last decade—a phenomenal rise in the use of electronic musical instruments, which synthesize sound digitally. A decade ago dedicated experimenters could coax musical sounds out of existing computers. Electronic analog sound synthesis systems, including electronic organs, were available, but they were limited and expensive; many tended to drift out of adjustment and go "out of tune."

Today we see a remarkable rise in commercial digital instruments, especially in popular music, and a decrease of the use of traditional acoustical instruments. Digital keyboards are cheaper than pianos, and their size suits today's cramped homes and apartments. Home computers produce primitive musical sounds, of increasing complexity. A studio full of advanced commercial digital equipment makes it possible for a skilled composer to produce a score for a film or a TV program without the expense of hiring an orchestra and rehearsal space. Singers, electric guitar virtuosos, and percussionists survive as essential, but, as John Appleton has observed, there are fewer jobs for trained instrumentalists. More and more music is played or produced digitally. Inspired by John Appleton's work at Dartmouth music departments and schools of music are responding by teaching students how to survive in our digital age.

Part of this training must involve studying the quality of the musical sounds produced by acoustic instruments. Much more than we might have expected, digitally generated sounds have imitated the sounds of traditional musical instruments. In part this may be because some digital systems, those based on *sampling*, start out with digital recordings of sounds played on acoustic instruments, or natural sounds, and modify these

180

sounds by transposition, the imposition of envelopes (how the sound rises and falls with time), and extension of the sound's length.

We still have a good deal to learn from traditional instrumental sounds. Through our culture, we have acquired a built-in acceptance for such sounds. They have been tuned to our ears, or our ears to them. We can learn a lot by studying them and trying to imitate them. Such study can teach us what is important to the ear in a sound wave, and what may not matter so much.

Here we encounter real puzzles. Surely the timbre of a sound must depend in some way on the spectrum of the sound, which we can measure with a commercial spectrum analyzer. Yet, musicians can recognize different instruments even when heard over a pocket-sized transistor radio that seriously distorts the spectrum, cutting off many low frequencies entirely. A saxophone still sounds like a saxophone when heard over such a radio, an oboe sounds like an oboe, a bassoon a bassoon, a violin a violin, and all are distinguishable from a French horn or the human voice.

The puzzle is great, for musicians can recognize musical instruments played over a wide range of pitch. Perhaps they learn to associate with one instrument certain qualities of sound throughout its pitch range. Even to the casual listener different pitches played on the same instrument have a

Figure 13-1 The percussion instruments of a large modern orchestra.

good deal in common. Dynamics are somewhat of a mystery. A *forte* sound is different from a *pianissimo* sound in more than intensity, just as a shout differs from a quiet voice. Merely turning the volume control up or down won't change one sound into the other.

The path to understanding is difficult. It must, as we have noted, involve some analysis of the sound, the sound waveform, or some other measurable quantity. But it must also involve synthesis. We can't be sure that we have "gotten at" the essential features of the sound until we can synthesize a satisfactory imitation. The imitation may, and usually should, leave out or simplify irrelevant features of the sound. Thus, if the sound includes a noisy component, any noise of the right spectrum and time envelope will do; we don't have to produce an exact replica of the original noise waveform.

Methods of Digital Synthesis

How we view a sound may depend on the means of synthesis we intend to use. In some methods of synthesis, an aspect of sound quality, such as harshness as opposed to mellowness, can be changed to some degree by adjusting one parameter. In other methods of synthesis, this may not be possible. Thus, in considering the quality or qualities of a sound, it is good to keep in mind various methods of digital synthesis.

When Max Mathews initiated computer synthesis of musical sounds in 1957, he chose to work with what is called *additive synthesis*. His computer synthesis programs, culminating in Music V (see Appendix F) are capable of additive synthesis—and much more. In additive synthesis we generate a sequence of individual sine waves whose amplitudes and phases change with time in any way we specify. Often, the phases are regarded as unimportant, and their changes with time are not controlled.

The virtue of additive synthesis is that in principle it can produce *any* sound. In practice, different means are provided for producing noise, which can be regarded as a huge number of sine waves spaced very close together in frequency. However, general additive synthesis is slow and "expensive." It gives us more control than we need to have over the spectrum of the sound wave. Further, we don't know what to do with this excessive control. Also, there is no inherent way to change the harmonic structure of the sound gradually in some sensible direction.

John Chowning's *fm synthesis* is an alternative to additive synthesis. In fm synthesis the frequency or phase of a sine wave is modulated or changed sinusoidally at the "original" (or carrier) frequency of the sine wave (or at an integer multiple of that frequency). The spectrum of the waveform thus produced will contain not only the frequency of the sine wave that has been modulated (the carrier frequency), but also harmonics of that frequency. By using fm synthesis we can get a sound consisting of a number of harmonics easily and cheaply. And we can increase the width of the spectrum (the number of harmonics that have appreciable amplitudes) by increasing the modulation index, that is, the amount of frequency modulation of the sine wave. Thus, changing one parameter, the index of modulation, can change the timbre of the sound generated gradually from simple to complex. The Yamaha DX7, the first inexpensive (around $2,000) digital keyboard, introduced in 1983, used fm synthesis.

An alternative to additive synthesis and fm synthesis is *subtractive synthesis*. In subtractive synthesis we generate a waveform rich in frequency components and put this through filters that emphasize certain ranges of frequency. In this way we easily get gross control over the shape of the spectrum without having to specify the amplitude of each harmonic. We can change timbre by altering the parameters of the filter, or filters, used to shape the spectrum. Subtractive synthesis has much in common with the functioning of acoustic musical instruments, including the human voice. In such instruments there is an initial periodic waveform with many harmonics — the vibration of the vocal folds (the voice), the vibration of the lips (brass instruments) or of a reed (woodwinds), the vibration of a bowed string (the violin family) or of a struck string (the piano). The spectrum of this initial sound is shaped by the resonances of the vocal tract, a column of air, or a soundboard before the sound is radiated. Subtractive synthesis, which emulates these features of acoustic sound production, is well adapted to producing sounds based on noise, including whispered speech, or in producing the noise component that must be added to pitched sounds that imitate some timbres.

Linear prediction, or the linear predictive vocoder, is a way of analyzing and reproducing the sound of a voice or a single instrument. In linear prediction we try to separate two aspects of vocal or instrumental sounds: the periodic excitation function and the overall spectrum. Thus, the reproduction of the sound is an instance of subtractive synthesis; the spectrum of a broad-band excitation function, which changes with time in pitch and somewhat in spectrum, is shaped by what amounts to a continually changing filter. Linear prediction has been used in modifying the pitch or the duration of the sounds of speech and of solo musical instruments. Paul Lansky has used the speech processing capabilities of linear prediction very effectively.

Chant synthesis, devised by Xavier Rodet, has been used very effectively in the synthesis of the singing voice. It has also been used in synthesizing a variety of other musical sounds. In Chant synthesis, a specified waveform is produced over and over again, in either an overlapping or nonoverlapping manner. Such a waveform may approximate the sound that the vocal tract produces when excited by a single glottal pulse. The rate of repetition gives the musical pitch. The shape of the repeated waveform, which usually rises fairly sharply and decays more slowly, gives the overall spectrum. The strength of Chant synthesis lies in the excellent results that have been attained. A problem is the production and use of a host of waveforms.

Finally, we can try to model the functioning of a musical instrument. In the very successful Karplus and Strong algorithm for duplicating the sound of a plucked string, the computer simulates the travel time of the wave back and forth along the string by means of delay, with a slight decrease in intensity for each traversal of the length of the string. The plucking is simulated by an initial short burst of noise that sets the wave in motion.

It has been at least thirty years since the human voice was first simulated by modeling waves traveling along the vocal tract, with reflections due to changes in the tract's diameter. Using the NeXT computer, Perry Cook at Stanford has shown that such simulations can now be carried out in real time.

Simulation of acoustic musical instruments, including the voice, is attractive because it may lead to control of sound quality, including the expressive features of performance, through varying a few physically meaningful parameters such as the excitation and configuration of the vocal tract, rather than through trying to duplicate directly the evolution of spectral features with time.

In addition to the generic approaches to synthesis discussed above, there are a number of complicated techniques used by Yamaha, Korg, E-mu, Ensoniq, Roland, and others that we might call hardware based. Powerful and flexible digital hardware makes it possible to combine and explore various ways of modifying sampled or synthesized waveforms. More hardware-based techniques are bound to appear in the future.

Whatever the means of synthesis, the importance of both analysis and synthesis in the study of musical sounds goes back at least as far as Helmholtz. His analytical tools were crude by our standards. He could watch the path traced out by a bright speck on a violin string or a tuning fork. He could seek out partials by listening successively through a number of glass resonators. He could listen for partials, carefully *hearing them out* by focusing his attention on a tone from a tuning fork or a musical instrument. He could calculate the relative strengths of the partials that are produced when a stretched string is plucked or struck.

Helmholtz's means for synthesizing sounds were even cruder. He could strike several tuning forks so they vibrated simultaneously. He could produce sounds on sirens with several different rings of holes. That was about all.

Helmholtz accomplished a great deal despite the limitations of the technology available to him. Yet he reached false conclusions. He believed that perception of musical pitch depends on the presence of the fundamental frequency. This is not true for tones toward the low end of the piano keyboard, or for orchestra chimes, or for bells. His other false conclusion was that the relative phases of sinusoidal components do not affect the timbre of a sound.

The invention of the vacuum tube and the emergence of analog electronics greatly helped such scientists as Harvey Fletcher in their study of musical sound. Nonetheless, limitations still remained. In this day of digital analysis and synthesis there are essentially no technological barriers. Alas, insight and technical expertise have scarcely kept up with the technology. Nevertheless, we have learned a great deal from Helmholtz's day to the present.

Fourier analysis (see Chapter 3) has taught us to think of the waveform of a periodic musical sound as the sum of sinusoidal components or partials of frequencies f (the lowest or fundamental frequency), $2f$ (the second harmonic), $3f$ (the third harmonic), and so on. Each frequency component or partial has a particular amplitude and phase. Figure 13-2 shows three different waveforms, each made up of the first sixteen harmonic partials with equal amplitudes. However, the relative phases of the partials are different in A, B, and C, and the waveforms are very different as well.

These three waveforms sound somewhat different for low fundamental frequencies f (pitches toward the lower end of the piano keyboard). Similar waveforms with more harmonic partials would sound very different for low pitches. But distinctions of sound due to different phases of the partials are lost as the pitch is increased. The ear simply does not have time to observe changes in the waveform during one period of the sound.

As we noted in Chapter 7, the basilar membrane of the ear performs a rough frequency analysis of a sound wave reaching the ear, but it does not measure phase directly. Sensation is influenced only by the relative phases of frequency components lying within *one critical bandwidth*, which we take as about a quarter of an octave, a minor third.

The first six harmonics of a musical sound span the interval two octaves plus a perfect fifth. Hence, their relative phase can't appreciably affect timbre, though difficult-to-hear phase effects can be synthesized in laboratory experiments.

Very high frequencies (for example, the seventh and higher harmonics) are essential to musical quality but their effects aren't very musically distinguishable. Suppose we say that the sixth harmonic of a tone has a frequency of 12,000 Hz and is musically important. The corresponding

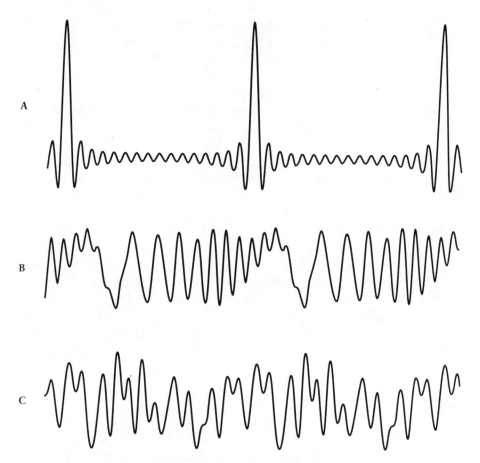

Figure 13-2 **A, B,** and **C** in this figure show two cycles or periods of waveforms with the same amplitude for each of sixteen successive harmonic partials, but with different phases for different partials. In **A** all harmonic partials peak at the same time, and the waveform is a sequence of pulses with wiggles between them. In **B** the phases are such that the higher-frequency harmonics are delayed with respect to the lower-frequency harmonics; during a period the frequency appears to rise, then to jump down at the end of the period and to rise again during the next period. In **C** the phases are random; the waveform is a short burst of noise that repeats over and over again. The lower the frequency (the longer the period) the more distinct the three waveforms sound. Toward the lower end of the piano keyboard **A** sounds clicky or pulselike, **B** sounds like a rapid succession of rising frequencies, and **C** sounds somewhat noisy. These distinctions can be made greater by including higher harmonics. As the fundamental frequency is increased, the distinction between the sounds of **A, B,** and **C** is gradually lost — pretty much so above middle C, and very much an octave higher. At high fundamental frequencies the ear doesn't have time to follow the changes in waveform within one period of the sound.

fundamental frequency is 2,000 Hz, roughly equivalent to a pitch of about C7, the next-to-highest C on the piano keyboard. Clearly, for this or higher pitches, the relative phase of the harmonics won't be important.

The fundamental frequency of the lowest tone on the piano keyboard, A_0, is 27.5 Hz. Its sixth harmonic has a frequency of 165 Hz. As we saw in Chapter 8, loudness of harmonics increases rapidly as frequency increases, especially in this region of the spectrum. If the higher harmonics of a 27.5 Hz tone have appreciable amplitudes, those above the sixth will certainly contribute to the sound as we hear it. Thus, at the low end of the piano keyboard the relative phases of such higher harmonics surely are important to timbre.

As shown in Figure 13-2, for fixed harmonic amplitudes, differences in the relative phases of harmonics can give waveforms that look like a sequence of sharp pulses at the fundamental frequency, or like a sequence of chirps (sounds of rising frequency) repeated at the fundamental frequency, or like a ragged section of noise waveform repeated at the fundamental frequency.

We can hear such differences in relative phase and waveform very clearly at the low end of the keyboard, but we cannot hear them at the high end. Roughly, distinctions in sound quality due to phase are clear up to around 200 Hz, about G_3, and vanish above about 400 Hz, about G_4, the G above middle C. We hear phase effects more easily with headphones than in more ordinary surroundings, as in a room.

Partials and Sound Quality

The partials of the sounds produced by acoustic musical instruments aren't quite steady in frequency and amplitude. Tiny, common variations in frequency (vibrato) and amplitude (tremolo) are important to a sense of a single, "natural" sound. For some instruments, notably the piano, the partials are not quite harmonic, as we discovered in Chapter 3. Both the stiffness and the tension of piano strings tend to keep them straight. The tension has the greater effect. Changes in tension are used in tuning. The smaller effect of the stiffness is greatest on the frequency of high partials, for which the string vibrates with many short bends. The extra force added by the stiffness of the string makes such partial frequencies appreciably higher than they would be for stretched strings without stiffness. The degree of *inharmonicity* is such that the fifteenth partial can have sixteen times the frequency of the fundamental, instead of fifteen times (a harmonic partial).

In 1962 Harvey Fletcher and his collaborators studied the sound of the piano by means of analysis and synthesis. They found that the warmth of the piano tone depends on this very inharmonicity of the partials.

Sounds synthesized with harmonic partials were bland and uninteresting. They had none of the wavering quality that gives the tone of the piano its warmth. The wavering quality is caused by the continual shifting of the relative phases of the higher partials.

The quality of the piano's tone must in some degree depend on what partials are actually excited when the hammer strikes the string. It is a commonplace of piano lore that the hammers are placed to strike at a point roughly one-seventh of the way along the string (see Figure 13-3), so that the hammer cannot set the string vibrating with the pattern of the seventh partial, shown in the drawing as a displacement of the string. But measurements of the spectra of piano tones show that the seventh harmonic *is* present.

Nonetheless, the hammer position must affect what partials will be excited, and in what relative intensities. The nature of the hammer is important, too. A hard hammer excites high partials and gives a bright tone. Softening the hammer by pricking the felt repeatedly with a special tool (in a process called "voicing") weakens the higher partials and gives a mellower tone.

The sound of a harpsichord is brighter than the sound of a piano. The plucking of the string by a small, sharp quill sets up many high-frequency partials, and these give the harpsichord its jangly sound.

The piano, the clavichord, the harpsichord, the guitar, the harp, and certain other stringed instruments have a characteristic plucked or struck sound. From the earliest days of computer sound synthesis, it has been observed than a short attack followed by a gradual decay, that is, any abrupt rise of amplitude coupled with any gradual decay gives this charac-

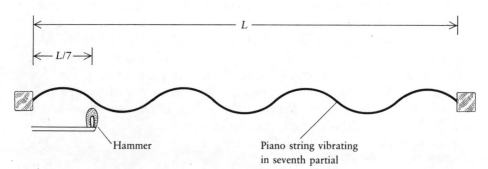

Hammer

Piano string vibrating
in seventh partial

Figure 13-3 In pianos, the hammers are placed about a seventh of the string's length from the end of the string. An argument can be made that this should prevent the excitation of the seventh harmonic in the vibrations of the string subsequent to striking. The argument must be fallacious, however, for piano tones do have a component of the seventh harmonic.

teristic sound, whatever the shape of the rise or decay, and regardless of waveform or spectrum.

The relative strengths of the various partials affect the quality of the sound. They help us to distinguish the sound of the harpsichord from the sound of the piano.

We noted in Chapter 5 that when many partials lie within one critical bandwidth, they give a sound a dissonant quality. As we saw, this is true of the sound of the harpsichord. It is also true of many early electronic sounds, and of some current electronic sounds.

If sounds don't have high partials, they sound dull. If they have many successive high partials, they sound buzzy. Some years ago I synthesized sounds that were bright but not buzzy by using high, *non*successive partials spaced at least a quarter octave apart. I found that Robert A. Moog of synthesizer fame has a patent on this excellent idea (U.S. patent 4,117,414, filed June 21, 1977).

The excitation of a violin string by bowing is quite different from excitation by striking or plucking. In the continually forced vibration of the string produced by bowing, the partials must be essentially harmonic, though they can waver a little in relative phase.

As observed by Helmholtz and confirmed by Mathews and others, the motion of the bowed violin string follows a sawtooth pattern, as shown in Figure 13-4. The bow drags the string along for a short distance, and then the string slips back, only to be picked up again and carried along with the motion of the bow. If this were the waveform of the sound of the violin, that sound would be harsh indeed. However, the soundboard of the violin has many resonances, which reinforce some partials and suppress others. Figure 13-5 shows a plot, against frequency, of the effectiveness of the violin body in transforming a partial of the string's vibration at the bridge into a partial of the sound wave that the violin produces. In the sound of the violin, a few partials whose frequencies lie close to the resonances of

Figure 13-4 The motion of a bowed violin string has a sort of sawtooth pattern of displacement with time. The bow drags the string along until a reflected wave from the *nut* (the small ridge at the peg-end of the violin, over which the strings pass) causes the string to slip past the bow. The bow catches the string again and pulls it along, as indicated in this figure.

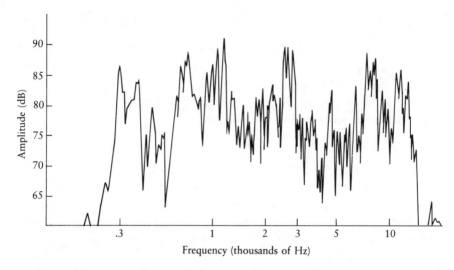

Frequency (thousands of Hz)

Figure 13-5 A sawtooth wave, such as that in Figure 13-4, has a very harsh sound. What we hear is not the motion of the string, but that motion translated into a sound wave in the air through the soundboard (and other structures) of the violin. The soundboard has many resonances. The curve in this figure shows how effectively a partial in the motion of the string is transferred to a partial in the sound wave radiated by the violin. Among the higher-frequency partials, some will lie near a resonant frequency of the violin and produce strong sounds; others will lie in the valleys of the curve and will produce little sound. This is very important to the sound quality of a violin. Much of the difference between good and bad violins depends on the location of the peaks and valleys of such a curve.

the violin are much more intense than those nearby. Jean-Claude Risset has noted that this has an interesting and important connection with vibrato.

The small variation of frequency causes amplitude modulation of the harmonics of the tone, because the vibrato shifts the frequencies of the harmonics toward or away from the resonant peaks shown in Figure 13-5. This effect is very apparent in some recordings of the waveforms of violin tones that Mark Dolson made in 1983 as part of his doctoral work at the California Institute of Technology. The waveform changed markedly during the vibrato, showing that the relative amplitudes of different harmonics do indeed change as the fundamental frequency of vibration of the string shifts up and down slightly.

Max Mathews and others have made electronic violins in which the sawtooth motion of the string is converted into an electric signal by means of piezoelectric material, which produces a voltage when stressed. In Mathew's early violins the signal was passed through an electrical network having seventeen to thirty-seven poles in order to simulate the resonances of the violin body. He later obtained a good "string" sound quality without such a filter. Perhaps the somewhat unsophisticated loudspeakers that he used had jagged resonances similar to those of an actual violin body.

In the violin the frequencies that are emphasized are determined by the resonances of the soundboard, not by the frequency of the note played. If a resonance at 2,200 Hz is excited by the tenth harmonic of a 220-Hz pitch, it will also be excited by the fifth harmonic of a 440-Hz pitch. In some other instruments, tone color depends instead on the relative intensities of particular harmonics. For example, closed organ pipes have a "hollow" sound because chiefly odd-numbered harmonics are present. Open pipes have a fuller sound because they include both even- and odd-numbered harmonics.

The Human Voice

The vowel sounds of the human voice are distinguished and perceived, regardless of pitch, because of the three chief resonances, or *formants*, of the vocal tract. Near the formant frequencies the intensities of the harmonics of the sound produced by the vocal folds are strong; harmonics far removed from the formant frequencies are weak. Indeed, we can distinguish the vowels in whispered speech, in which all frequencies are present, not just a sequence of harmonics. In part A of Figure 13-6 we see the spectrum of a whispered *a* as in *had*. In part B we see the same vowel at a pitch of 200 Hz. In part C we see the same vowel at a pitch of 400 Hz. In parts B and C the envelope, or outline of the peaks of the spectrum, is roughly the same as for the whispered vowel. This envelope represents the resonances of the vocal tract.

I have said that the frequencies of the formants characterize a vowel sound in speech, regardless of pitch. This is not quite true for the singing voice, as Johan Sundberg has shown. For one thing, the singer manages to produce a high *singer's formant*, which makes the singing voice intense in a frequency range in which common orchestral sounds have little power (see Figure 9-3). Furthermore, sopranos shift their formants when they sing very high notes. This both makes the sound louder and alters its quality. Computer synthesis has shown that if sopranos do not shift their formants, the voice sounds like that of a child, not a woman. Finally, some singers can control the formants so that they coincide with particular harmonics. They can seem to sing several notes at once, though all are, of course, harmonics of the frequency of vibration of the vocal folds. The accurate control of formant frequencies gives a wonderful quality, not found in normal speech or singing, to some Buddhist chanting.

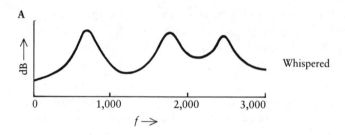

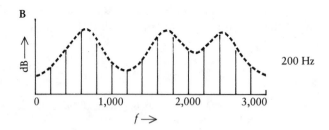

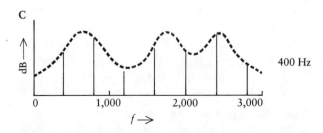

Figure 13-6 Vowel sounds are characterized by resonances of the vocal tracts, or *formants*, which make the speech sound more intense in narrow frequency ranges. The curves shown in this figure illustrate the effect of these resonances for the vowel *a* as in *had*. Curve **A** illustrates a whispered *had*. This is a noiselike sound, and a whole range of frequencies is present. *Had* spoken with a pitch frequency of 200 Hz is illustrated in part **B**. Only partials of frequencies 200 Hz and its harmonics are present. The dashed *envelope* curve indicates the effects of the resonances of the vocal tract on the intensities of these partials. In part **C**, the pitch frequency is 400 Hz. There are only half as many partials as in part **B**. One might guess that it is harder to find the formants from a recording of a female voice than from a recording of a male voice, and this is true. The widely spaced partials of the female voice do not indicate the formant frequencies very clearly.

The formant frequencies are clearly an extremely important aspect of the human voice and are worthy of our attention. Figure 13-7 shows vocal tract configurations, and Table 13-1 lists the corresponding formant frequencies for common English vowels.

The first two formants are most important to the identification of vowel sounds. In Figure 13-8 the logarithm of the frequency of the second formant is plotted against the frequency of the first formant for the vowels listed in Table 13-1. The length of the sides of the little square in the figure shows the change in position corresponding to a frequency shift of one semitone in the frequency of either the first or second formant. Because the logarithm of the frequency is used rather than the frequency itself, shifting the frequency of the first or second formant by a given number of semitones moves the vowel sound the same distance in the plot, whatever the formant frequency may be.

We see from Figure 13-8 that the formant positions are several semitones apart. We aren't as acutely aware of a semitone shift in the resonant frequency corresponding to a formant as we are of a semitone change in pitch.

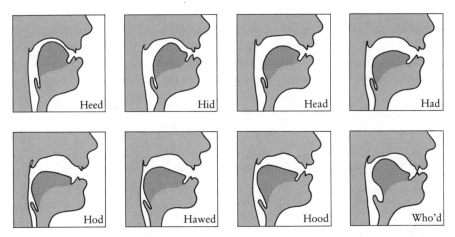

Figure 13-7 Vocal tract positions for some English vowels. The vowels in *heed, hid, head,* and *had* are called *front vowels,* because the highest point of the tongue is in the front of the mouth. The vowels in *hod, hawed, hood* and *who'd* are called *back vowels,* because the highest point of the tongue is in the back of the mouth. The tongue is highest in the vowels of *heed* and *who'd,* which are called *high* or *close* vowels, and lowest in the vowels of *had* and *hod,* which are called *low* or *open* vowels. As to timbre, the vowel of *heed* seems shrill, and the timbre of the vowel in *who'd* seems low or dull.

Table 13-1 Formant Frequencies of Common Vowels.

	Heed	Hid	Head	Had	Hod	Hawed	Hood	Who'd	Hud	Heard
f_1	270	390	530	660	730	570	440	300	640	490
f_2	2,290	1,990	1,840	1,720	1,090	840	1,020	870	1,190	1,350
f_3	3,010	2,550	2,480	2,410	2,440	2,410	2,240	2,240	2,390	1,690

The many successive black or white keys on the piano keyboard are separated by semitones. There are only a few vowels, each characterized by its resonances or formants. When different people utter the same vowel, their formants are somewhat different. If the formants of different vowels were too close together in frequency, it might be difficult to understand different speakers. Whatever the explanation may be, in the perception of

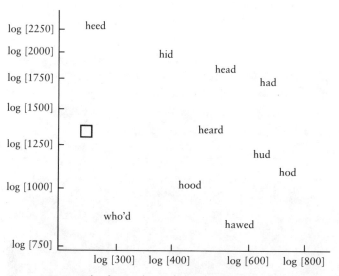

Figure 13-8 The logarithm of the frequency of the second formant is plotted against the logarithm of the frequency of the first formant for the vowels listed in Table 13-1. The length of the sides of the little square in the figure shows the change in position corresponding to a frequency shift of one semitone in the frequency of either the first or second formant. Because the logarithm of the frequency is used rather than the frequency itself, shifting the frequency of the first or second formant by a given number of semitones moves the vowel sound the same distance in the plot, whatever the formant frequency may be. (Based on work of David Mettinger at Stanford.)

pitch we make considerably finer frequency distinctions than we make in the perception of vowels.

Vowels, with their formant frequencies, can be synthesized in various ways. Straight additive synthesis is possible. We can simply specify the amplitudes and frequencies of all harmonics present, but this seems rather artificial.

John Chowning has produced good sung vowels by means of fm synthesis. The nominal frequencies of all formants are replaced by nearby frequencies that are harmonics of the fundamental frequency of the vowel to be synthesized. Then a sine wave is produced at each of these harmonic frequencies, and each of these sine waves is frequency modulated by a sine wave of the fundamental or pitch frequency. This produces a spectrum with peaks at the (slightly altered) formant frequencies. The spectrum is made up of harmonics of the fundamental frequency. It falls off in amplitude away from the shifted formant frequencies.

Chowning discovered a remarkable phenomenon. Though the spectrum so generated was "right," the ear did not hear the frequencies as a single, vowel-like sound — that is, not until a common vibrato was applied to all of the frequency components.

The Chant method of synthesis, noted earlier, is particularly appropriate to vowel synthesis. We can think of a vowel sound as generated by successive puffs of air passing the vocal folds. Each puff excites brief oscillations in the vocal tract, which subsequently die out. In Chant synthesis, a waveform is generated which corresponds to the sound produced when one puff excites the vocal tract. The overall waveform for a vowel sound is produced by adding a succession of such waveforms, produced regularly (or with vibrato) at the pitch frequency.

Chant synthesis is a more "natural" approach than synthesis of vowels by straight additive synthesis or by fm. Subtractive synthesis, in which successive "puff" waveforms are passed through a filter with a passband shape given by the formant resonances, seems still more natural; it corresponds more closely to the actual production of speech and song. Wouldn't it be still more "natural" to simulate the actual structure of the vocal tract, the propagation of waves along a tube with various constrictions corresponding to those shown in Figure 13-7? This is an old idea, but until recently it could not be done in real time. As we noted earlier in this chapter, Perry Cook has succeeded in synthesizing speech and song sounds in real time using a NeXT computer.

The excitation (sound pressure and sound velocity) at any point along the vocal tract can be represented as the sum of a rightward-traveling wave and a leftward-traveling wave, each with an independent amplitude or power. The effect of a constriction of the vocal tract is to transfer power between the leftward-traveling wave and the rightward-traveling wave, with no net change in the total power of both waves. Perry Cook simulates the waves and their reflections using time delay in representing the travel of the wave from one reflecting discontinuity to the next. The shape of the vocal tract is represented by means of five adjustable intermediate reflecting discontinuities, with fixed reflection at the lips and the vocal folds.

This simulation of the vocal tract produces good spoken and sung vowels. Through motions of the tract shape, consonants can be produced. In speech or song, the hissing or noisy part of consonants is produced by airflow through constrictions (blow through your lips or past your tongue to demonstrate this). Cook introduces noise of the proper spectrum at reflecting discontinuities, noise whose intensity varies properly throughout each period and changes properly with the degree of closure of the vocal tract.

Of all the available means for synthesizing speech sounds, such simulations of the vocal tract seem most appealing. They allow us to mimic various features of actual speech sounds. Let us, then, consider a very simple feature of vowel sounds.

We might conclude that, however we produce a vowel sound, frequency spectrum is all there is to the sound. Try saying *ah* in a very prolonged monotone, without any vibrato or change in pitch or intensity. I think that you will find that the sound loses something of its *ah* character. It becomes a rather buzzy, not too pleasant tone, without much character. What happened to its *ah*ness?

When we hear an absolutely steady vowel sound, our first impression is to recognize what vowel we hear. As the vowel is prolonged, we come to hear it more as a buzzy sound, perhaps because much of our nervous system is constructed to respond to changes rather than to a continuing, unchanging stimulus. We sense the onset and the first parts of sounds differently from their later parts. And throughout the sound, changes are welcome. The rise and fall of sound, vibrato, its onset and diminution — all are important to the ear. Indeed, in a passage of music it is important that successive tones don't sound *exactly* alike.

Physical Aspects of Musical Timbre

Time variations can be crucial to the very nature of musical sounds. This was demonstrated unequivocally in experiments carried out by Jean-Claude Risset at Bell Laboratories in the 1960s. Risset used a computer to analyze in great detail the rise, fall, and variation with time of the various partials of recordings of short trumpet tones played by a professional trumpeter. He found the sounds to be exceedingly complex. By using the computer to synthesize sounds that matched selected fine details of the analyzed trumpet sounds, he found that some complex features of the real trumpet sounds were important to the ear, and some weren't. For example, short-term fluctuations of the amplitudes of various partials turned out not to be important to the ear. Neither was the short burst of noise found at the beginning of real trumpet tones.

What did prove important was that the higher partials start later and fall earlier than the lower partials. Although random variations in amplitude of the partials proved unimportant to the ear, random variations of the frequencies of the partials were important in giving the synthesized sounds the brassy sound of a real trumpet. Omission of all frequencies above 4,000 Hz, from either real or synthesized sounds, did not make them less recognizable as being trumpet sounds, but did make them seem less brilliant. When trumpet sounds were synthesized with reasonable attention to the rise and fall of the intensities of various partials, and with appropriate random frequency changes of vibratos of the partials, trumpet players could not tell the synthesized short trumpet sounds from real trumpet sounds.

Dexter Morrill and others have since synthesized excellent trumpet passages. What is needed to synthesize trumpet tones is a progressive change in the spectrum, that is, a progressive change in the strength of higher harmonics. By using fm synthesis we can get this by simply changing the frequency deviation, the intensity of fm-ing. Thus, it is simple to realize an important ingredient of brassy sound with fm synthesis.

In his thesis, written at Stanford in 1975, John M. Grey set out to explore, compare, and differentiate the sounds of a variety of musical instruments through the method of analysis by synthesis. He identified and eliminated those features of real musical sounds that make little or no difference to the trained ear. He then synthesized instrumental sounds equal in duration, loudness, and pitch that proved to be difficult or impossible to distinguish from sounds produced by the instruments they imitated.

Grey asked trained musicians to rate the similarity of pairs of musical sounds. The rating scale was *very dissimilar*, 1 to 10; *dissimilar*, 11 to 20; *very similar*, 21 to 30. He sought to get a simple representation of the similarities of these sounds by means of a computer analytical technique called *multidimensional scaling*. This gave him the three-dimensional representation shown in Figure 13-9. In this representation, each cube or square stands for a particular instrument, and distance between cubes or squares measures similarity — the shorter the distance, the greater the similarity.

On the basis of similarity, Grey found that instruments fell into three families, each with several subfamilies. These are as follows:

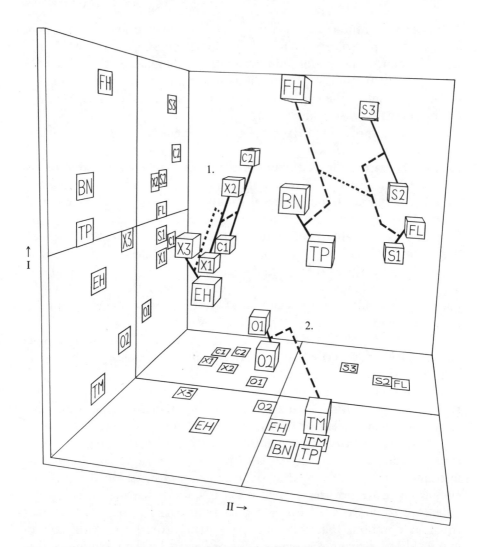

Figure 13-9 Three-dimensional display of differences and similarities between instrumental sounds based on numerical ratings of similarity or dissimilarity for various pairs of sounds. The display was obtained by a computer technique called *multidimensional scaling*. In it, instrumental sounds that were judged to be similar are close together, those judged dissimilar are far apart. The dotted and dashed lines that connect the members of several groups of instruments were obtained by a different technique, called *clustering*. The squares on the walls are two-dimensional projections of the cubes, showing their positions up-and-down and forward-and-back (left wall), and left-and-right and forward-and-back (bottom wall). The abbreviations are: O1, O2, oboes; C1, C2, clarinets; X1, X2, X3, saxophones; EH, English horn; FH, French horn; S1, S2, S3, strings; TP, trumpet; TM, trombone, FL, flute; BN, bassoon.

1. E-flat clarinet (C1); soprano saxophone *mf* (X1); soprano saxophone, *f* (X3); bass clarinet (C2); soprano saxophone, *p* (X2); English horn (EH).
2. oboe (O1); muted trombone (TM).
3. bassoon (BN); French horn (FH); cello, *sul ponticello* (bowed near the bridge) (S1); cello, normal (S2); trumpet (TP); flute (F1); cello, *sul tasto* (bowed near the fingerboard) (S3).

Grey next looked for the physical characteristics of the sounds responsible for these similarities. He found that up-and-down (I) can be interpreted as *spectral energy distribution*. Near the top (FH and S3), the spectrum is narrow and has its peak at a comparatively low frequency. Near the bottom (TM), the spectrum is wide and peaks at a higher frequency. TP and X3 lie in midrange.

Left-and-right (II) seems to depend on whether the partials rise and fall at the same time. In the sounds of the woodwinds (left), the various partials do rise and fall at the same time. This is not so for strings, flute, or brass, nor for the bassoon, the most complicated of the woodwind instruments.

Forward-and-back (III) seems related to the initiation of the sound. In the sound of the flute (FL) and strings (S1, S2, S3), a short burst of noise precedes the tone and is important for its quality. This burst of noise is not so important for the trumpet (TP), the trombone (TM), or the bassoon (BN). However, too large a burst of noise at the beginning of a synthesized violin sound gives the effect of an inexpert player.

It is not entirely clear just what is represented by the up-and-down spectral energy dimension, I. Perhaps this is a sort of average of the frequencies of the partials, weighted by their computed loudness.

It is clear that many physical aspects of musical sounds, including spectral information, contribute to their timbres. (Spectral information also enables us to distinguish the various vowel sounds.) The relative times and rates at which various partials rise is important in brass sounds. An initial high-frequency burst of noise is more essential to the timbre of some instruments, such as flutes and strings, than to others, such as the trumpet. And we should remember that a fast attack and a gradual decay gives the effect of a plucked or struck string.

Bells, Gongs, and Chimes

Bells and gongs are also struck. Bells and gongs differ from stringed instruments in that any harmonic relation among some of the partials can be obtained only by deliberate design. Without a harmonic relation among partials, tones of bells and gongs do not give conventional effects of consonance and dissonance, even if recognizable melodies can be played on them.

A great deal of study has been devoted to bells, gongs, and related instruments. Among these are orchestral chimes, which are long, uniform metal tubes of various lengths that are suspended from one end so that they can flex and vibrate freely (see Figure 13-10). The frequencies of the first few partials of a typical chime are shown in Table 13-2. Here the fourth, fifth, sixth, and seventh partials are approximate harmonics of a frequency equal to $4.5 f_0$, which is therefore heard as the perceived pitch of the chime. This is a manifestation of the residue, or virtual, pitch (see Chapter 6).

We have all observed that a thin board bends more easily than a thick board. Imagine that the tube of a chime is not exactly round, but is flattened a little. It will bend more readily in the flattened direction than in

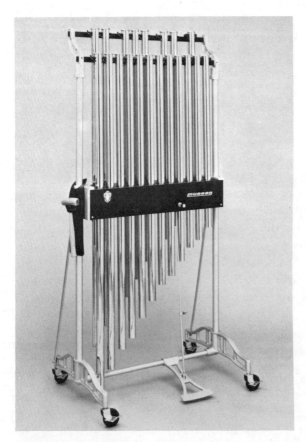

Figure 13-10 Orchestral chimes.

Table 13-2 Partials of a Chime.

1st	2nd	3rd	4th	5th	6th	7th
f_0	$2.76\,f_0$	$5.40\,f_0$	$8.93\,f_0$	$13.34\,f_0$	$18.64\,f_0$	$31.87\,f_0$
—	—	$(4.5\,f_0)$	$2 \times 4.47\,f_0$	$3 \times 4.45\,f_0$	$4 \times 4.66\,f_0$	$7 \times 4.55\,f_0$

the thickened direction. A chime that is very slightly out of round will have two sets of partials, one with slightly higher frequencies than the other. These partials will beat, giving a slightly wavering sound. This beating or wavering is undesirable in chimes, but is an attractive feature of the sound of gongs. Such wavering has been used effectively in computer-synthesized bell and gong tones.

The glockenspiel consists of loosely mounted metal bars. These have many high partials, some corresponding to bending and some to twisting motion. The high partials die away rapidly after the bar is struck. The perceived pitch is the frequency of the lowest, bending partial. The glockenspiel gives high-pitched sounds that are written two octaves lower than they sound.

The xylophone (see Figure 13-11) and the marimba (see Figure 13-12) use wooden fiberglass bars, thinner in the middle than at the ends. This thinness of the middle of the bar causes the second partial of the marimba bar to lie two octaves above the first partial. The xylophone bar is thinned less, and so the second partial has a frequency three times that of the first partial. Closed tubular resonators under the bars of marimbas and xylophones increase the sound intensity of the first partial and make it die out more quickly. In the xylophone, the closed tube under a bar has another resonance, at three times the frequency of the fundamental — that is, at the frequency of the second partial of the xylophone bar — and so strengthens its intensity. This is one reason why the xylophone has a brighter quality than the marimba. However, the xylophone also sounds brighter because it is commonly played with hard mallets, whereas the marimba is more often played with soft mallets.

The vibraphone, vibraharp, or "vibes" has aluminum bars shaped much like those of the marimba. The vibrations of these bars decay more slowly than those of the marimba. Motor-driven disks between the tops of the resonators and the bars above them alternately open and close a passage between the bars and the resonators, and give the vibraphone sound its wavering quality.

The phenomenon of residue, or virtual, pitch also explains the perceived pitch of bells, whose modes of oscillation are very complex. The

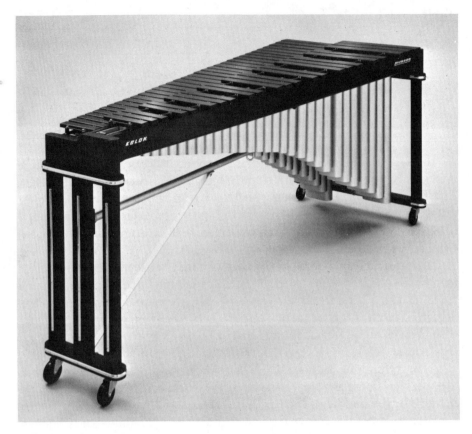

Figure 13-11   A xylophone.

bells of carillons are ingeniously designed to approximate closely a prescribed set of partials. Around 1644, Jacob van Eyck, a Dutch composer
and the carilloneur of the main church of Utrecht, asked the brothers
Frans and Pieter Hemony how to tune the lower five partials of bells to the
frequency ratios 1:2:2.4:3:4, in order to obtain the best-sounding carillon. The Hemony brothers succeeded and became the most famous bell
founders in history. Figure 13-13 shows a Hemony bell.

Table 13-3 shows and describes the partials of the Hemony, or *minor
third*, bell. We can observe that the significant partials are separated by
musical intervals, and that the prime, third, and fifth form a minor triad.
We can also see that except for the third, the frequencies of all the partials
listed are harmonics of the hum tone, the lowest frequency present, an
octave below the prime.

Figure 13-12 A marimba.

If the third were a major third rather than a minor third above the prime, then the prime, third, and fifth would form a major triad. We can conclude that all the listed partials would therefore be harmonics of a tone one octave below the hum tone, or two octaves below the prime. The idea of carillon bells with a major rather than a minor third goes back to the early part of this century, but bell founders found it impossible to realize this objective with bells of conventional shape.

In 1987 a collaborative publication of a realization of major third bells was produced by ingenious people at the IPO (Institute for Perception Research) at Eindhoven and the Royal Eijsbouts Bell Foundry at Asten, both in the Netherlands. The program of work included an investigation of listener reaction to synthesized minor and major third bells, detailed computer calculations investigating and optimizing the modes of bells of unconventional shapes, the casting and refining bells of a new shape, and finally the construction of a four-octave transportable carillon. Figure 13-14 compares the shape of the major third bell with that of a minor third bell.

Figure 13-13   A Hemony bell.

Table 13-3 Partials of the Hemony, or Minor Third, Bell.

Partial	Relation to f_p, the Perceived Pitch
Hum tone	0.5 f_p an octave down)
Prime	f_p
Third	1.2 f_p (a minor third up)
Fifth	1.5 f_p (a fifth up)
Octave	2 f_p (an octave up)
Upper third	2.5 f_p (an octave plus a major third up)
Upper fifth	3 f_p (an octave plus a fifth up)

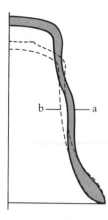

Figure 13-14 The major-third (a) versus the minor third bell (b).

The major third bell sounds less dissonant and has a shorter decay time than conventional bells. Successive chords overlap less in time. Lay listeners prefer it. Musicians find it better adapted to the major scale. However, the reactions of carilloneurs vary greatly and have been mostly negative.

This endeavor was a truly marvelous wedding of science, engineering, and art. Its success is deserved.

Drums

Kettledrums, or tympani, have many partials, but the pitch is that of a mode of vibration called the *principal tone*. Several partials present whose frequencies are two, three, four, and five times *half* the frequency of the principal tone. For some unexplained reason the pitch of the kettledrum corresponds to the frequency of the principal tone, instead of being an octave lower. Drum makers of India have ingeniously made nonuniform drumheads that vibrate with almost harmonic frequencies and give a clear sense of pitch (see Figure 13-15). But most drums, such as bass drums, snare drums, tomtoms, conga, or bongos, have no clear pitch and are used solely for their percussive effect.

Analysis and Synthesis of Musical Sounds

In the foregoing discussions we have considered both the acoustical qualities of musical instruments and problems of analyzing and synthesizing the sounds that they produce. The general intent has been to fit together various aspects of analysis and synthesis. Indeed, can there be, is there a unified method of analysis and synthesis that is well suited to all musical

Figure 13-15 An Indian kettledrum

sounds and will give an apt description of them? There may be, in work done at Stanford by Xavier Serra and described in his doctoral dissertation in 1989.

As we noted in Chapter 3, any waveform can be represented by a sequence of spectrograms of successive overlapping portions of the waveform; the original waveform can be reconstructed accurately from such a sequence of spectrograms. Each such spectrogram tells how amplitude and phase vary as a function of frequency. This process of analysis and resynthesis has been called a *phase vocoder*. Serra made use of this process to his ends, taking successive spectra at intervals of around 10 milliseconds.

Such successive spectra do not in themselves give a deep insight into musical sounds. Serra's innovation was to use successive spectra in dividing the signal into two parts—a *deterministic*, or predictable part, and a *stochastic*, or unpredictable, noisy part. The deterministic part Serra took to be clear peaks which in several successive spectra change just a little in amplitude and phase. This part of the spectrum Serra resynthesized by generating the individual sinusoidal components whose amplitudes, frequencies, and phases changed with time in the fashion indicated by the successive spectra.

Serra then subtracted each spectrum of this waveform from that of the total waveform. This left those spectral lines which appeared and disappeared within a few spectra, and the part of the spectra that showed no clear line structure. Serra considered this remaining portion of the spectrum to be the stochastic or unpredictable part. Further, he replaced this part of the spectrum with a noise that had roughly the same overall spectrum as the stochastic part but that didn't match it in waveform.

Serra tested this division of the signal into a deterministic and stochastic part and his method of resynthesizing musical sounds by listening separately to the deterministic and stochastic parts, and then adding them and listening to their sum. A piano sound reconstructed from the deterministic spectra alone didn't sound like a piano. With the stochastic or noise portions added, it sounded just like a piano. The same was true for a guitar, a flute, a drum, even the human voice. The relative power in the stochastic noise spectrum varied from sound to sound, and, for a given sound, the relative noise power varied with time. During the piano attack, for example, the noise power was large, and fell after the hammer left the string.

Serra used this approach successfully in modifying musical sounds. In principle, such a combination of noise and deterministic sound could be used to construct new, "natural-sounding," appealing instrumental sounds. In practice, it isn't easy to fit a noise component and a deterministic or tonal portion together so as to get a unified, instrumentlike sound. This is reminiscent of a problem we encountered earlier.

In Chapter 6, I described computer-generated sounds with nonharmonic partials that didn't "hang together." The upper partials didn't fuse with the lower partials to give a sense of a single musical sound. Nevertheless, we hear bells, gongs, drums, and even wood striking wood (as in castanets) as single, distinguishable sounds, even though their partials are nonharmonic.

In part, a sharp attack followed by a slow decay helps to make sounds hang together. During doctoral work completed at Stanford in 1980, Elizabeth Cohen found this to be true for stretched tones, unless they were too strongly stretched.

Perhaps in part we identify some complex natural sounds as single sounds rather than mixtures because we have heard them many times, and have learned to recognize and name the physical source. Certainly, by careful listening we can hear various frequency components of sounds such as those of chimes, bells, and gongs, yet this does not keep us from identifying the whole sound as one sound, characteristic of that particular instrument. It may be that we have not yet identified some subtle characteristic by which we are able to recognize certain kinds of sounds as sounds from a single source.

14 Perception, Illusion, and Effect

*T*he human organism, ill knowing and ill understood, confronts itself across a perplexing territory of which it has only a rudimentary map. A listener is endowed with, and bound by, abilities of perception, keen to detect some subtleties, deaf to others, and open to deceit and illusion. The composer of traditional music has a canny knowledge of the sounds that players can evoke from instruments and of how these sounds can be made to contrast or blend. A composer of digital sounds knows something of how and why the paper cone of a loudspeaker can ring like a bell, blare like a trumpet, speak with a human voice, or produce sounds and illusions never before heard by human beings.

Our constitution, our mind, and our senses are central to the overall process of the production and perception of music. There is much that we do not know. Indeed, recent discoveries make us appreciate the depth of our ignorance.

Medical literature gives curious accounts of the *amusias*, defects of musical ability associated with disease or injury of the brain. These include loss of ability to sing (without words), loss of ability to write musical notation, loss of ability to recognize familiar melodies, and loss of ability to read musical notation. There have been studies of the impairment of more detailed skills, such as the correct perception of temporal order, of simultaneity, of duration, and of rhythm. Some of the deficiencies are associated with impaired function of the left, or "dominant," hemisphere of the brain, where, in right-handed people, speech and the understanding of language reside. But musical deficiencies and speech disorders or *aphasias* are not inevitably linked.

Although right-handed people who have suffered extensive damage to the left hemisphere of the brain may be able to speak and understand only with great disability, or not at all, work by Roger Sperry and his followers on split-brain patients, in whom the two hemispheres have been surgically isolated, has broadened and refined our understanding of speech, understanding, and other abilities. The right hemisphere can exhibit rudimentary language capabilities. It is better than the left at solving difficult geometric puzzles. The relative predominance of various abilities in the two hemispheres is clearly a fact, but not one yet completely understood.

There has been a good deal of controversy regarding the loss of musical abilities. It has been known for many years that injuries to the left hemisphere *can* result in loss of ability to sing or whistle a tune; yet in 1966 a patient whose entire left hemisphere had been removed, and who had lost the ability to speak, was found to be able to sing familiar songs with few articulatory errors.

Such puzzling phenomena probably indicate that a complex combination of abilities is needed for the perception and performance of music. Recently, it has become possible to locate the site of mental activities by injecting into a subject's arteries a radioactive substance that is absorbed by the part of the brain engaged in a task. Several subjects were asked to report whether two groups of tones were different or similar. In some subjects, those who tried to remember the succession of tones as a heard melody, the right hemisphere was more active during the task. In other subjects, those who mentally plotted the tones on a musical staff, the left hemisphere was more active. Here was one simple musical task. Different people approached it differently, with different parts of the brain. More recently, positron emission tomography (PET) has been used to pinpoint changes in cerebral blood flow to an accuracy of a few millimeters. Increased blood flow has been observed in different cerebral regions when a subject's attention was directed to changes in shape, color, or velocity in a visual stimulus. Similar studies have been made in listening to language; the cerebral response to an unknown tongue lies in a different region from that to a known tongue. How we would like to have such maps of the brain's response to music, or to attention to music!

Perception of Musical Illusions

We don't really know much about the perception of musical sound, except that it can be very complex and can differ from person to person.

Musical illusions are interesting in themselves and can shed light on our complex powers of perception. In the October 1975 issue of *Scientific American*, Diana Deutsch published an interesting article on musical illusions. One of these illusions is represented in Figure 14-1. As part A shows,

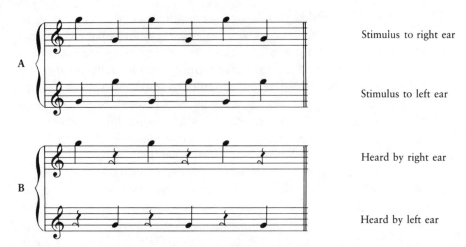

Figure 14-1 A musical illusion described by Diana Deutsch. The stimuli presented to the left and right ears through headphones are shown in part **A,** and what the subject heard is shown in part **B**. Although both high tones and low tones were presented to each ear, the right ear heard only high tones, the left ear only low tones. (Some people do not experience this illusion.)

a sequence of paired tones was presented to the right and left ears: first high to the right, low to the left; then low to the right, high to the left; and so on. As part B shows, a right-handed listener commonly heard a high tone in the right ear, followed by a low tone in the left ear, then a high tone in the right ear, and so on. Not only did the right ear disregard the low tones, and the left ear the high tones, but the left ear "heard" a low tone that was present only to the right ear. Deutsch suggested that only one ear (in this case, the right) perceived pitch and that localization was perceived by a separate mechanism that homed in on the higher pitch. Another illusion described by Deutsch, one that may seem congenial to musicians because, in a sense, it brings order out of seeming chaos, is shown in Figure 14-2.

Whatever the explanation of such phenomena of perception, they should warn the computer musician to beware. However, such binaural presentations of sounds are unusual in music. Other peculiar effects occur more commonly.

Some effects can be studied most easily by the means of computer-generated sounds, whose qualities, including timbre, can be specified and varied with great accuracy. Figure 14-3 illustrates an experiment described

A

B

Stimulus to
right ear

Stimulus to
left ear

Heard by
right ear

Heard by
left ear

Figure 14-2 Another musical illusion described by Diana Deutsch. The tones presented to the left and right ears, in part **A**, make up an ascending scale and a descending scale. What was heard is shown in part **B**. The right ear heard the higher tones as running partway down the scale and then up again. The left ear heard the lower tones as running partway up the scale and then down again.

in 1978 by David L. Wessel. Staff A shows the pitches of successive notes. The timbres of the notes marked + and × can be similar or very different.

When all notes were the same, or only slightly different, in timbre, the succession of notes was heard as written in part A; that is, as a repeated pattern of three ascending notes. However, as the difference in timbre between + and × increased, the notes were heard as two separate voices (as shown in part B), each voice repeated pattern of three descending tones. Extensive studies of such *streaming* effects by Albert S. Bregman, Wessel, and others show that notes can be bound together in a stream by differences in spectrum, but not by differences in attack. Those interested in pursuing these and related matters further should consult Bregman's book *Auditory Scene Analysis* (MIT Press, 1990).

This powerful effect of timbre that Wessel found led him to suggest that timbre could be used as a musical "dimension," like pitch or loudness. His work, and earlier work of John M. Grey (see Chapter 13), make it possible to map at least some timbres in an orderly way, as in the three-dimensional timbre space of the diagram in Figure 13-9. Can we make music out of an organized progression through such a space? Timbral transitions have been used effectively in Chowning's *Turenas* and *Phobē*,

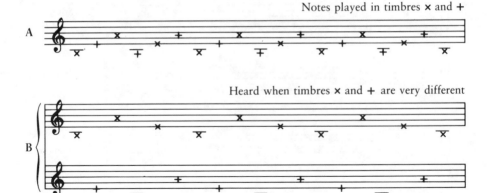

Figure 14-3 A musical illusion described by David Wessel. The notes shown by x and by + in part A may be played with the same or with different timbres. When played with the same timbre (as in part **A**), they are heard as sequences of three ascending notes. When the timbres differ enough (as in part **B**), two voices are heard, each consisting of a repeated pattern of three *descending* tones.

and in several compositions by Jean-Claude Risset. Many of Chowning's and Risset's pieces are now available on compact discs issued by Wergo.

The effect shown in Figure 14-3 — producing more than one voice by playing successive notes with different timbres or different ranges of pitch — is not uncommon in traditional music. It is used to particularly good effect in Johann Sebastian Bach's works for unaccompanied violin (for example, see the score in Figure 14-4). Separation in pitch helps to distinguish the voices (in accordance with Deutsch's observations), but a good violinist will use distinctions in timbre as well to help separate successive notes into two voices, in accord with Wessel's results.

The foregoing observations indicate that our hearing tends to associate tones of like timbre and to avoid perceived jumps in pitch. A good deal of recent music has made use of widely separated pitches in a single melodic line, and some composers have specified that successive pitches of a melodic line be played by different instruments. Both practices are hard for performers. In traditional music, large leaps or repeated leaps are used in order to obtain particular effects, such as a dramatic effect in Mozart's great arias for the Queen of the Night in *The Magic Flute*, or the effect of a bugle call in *"Non più andrai"* in *The Marriage of Figaro*. Differences of timbre have commonly been used to *distinguish* one musical line from another.

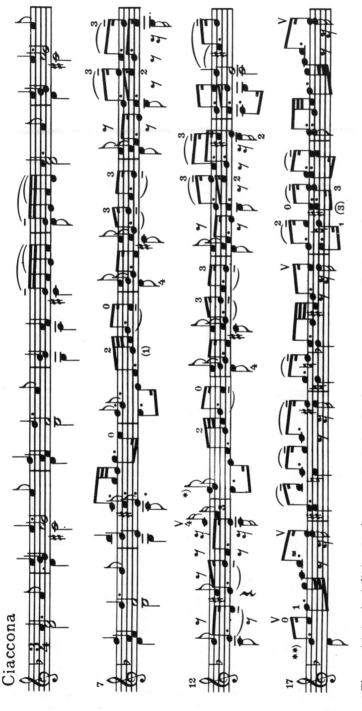

Figure 14-4 A violinist can bow at most two simultaneous notes; yet the beginning of this chaconne by Johann Sebastian Bach, from his Partita No. 2 for unaccompanied violin, calls for several four-note chords. Such a chord must be approximated by playing the notes *arpeggio*, in sequence. Furthermore, starting at bar 10, three distinct voices are indicated. The ear can separate these only because the voices move up and down in small intervals and don't overlap in pitch, though any aid that the violinist supplies by differences in loudness or timbre will help.

Because of its flexibility and accuracy, the computer can be used to produce auditory illusions that would be difficult or impossible to attain in any other way. In one very striking illusion produced by Jean-Claude Risset, the pitch of a sound recorded on tape *falls* slightly when the tape speed is doubled, going from 3.75 inches a second to 7.5 inches a second. Clearly, all sinusoidal components have doubled in frequency. Why has the pitch gone down?

Figure 14-5 shows how this is done. The frequencies of the partials of the tones are plotted as vertical lines on an octave scale, and their intensities are shown as the heights of the lines. The frequencies and intensities of the partials for the tape played at 3.75 inches a second are shown in part A of the figure. All partials are separated by 1.1 octaves. When the tape is played at 7.5 inches a second, the frequency of each partial is doubled and therefore shifted up an octave. The new first partial has a frequency about a tenth of an octave below that which the second partial formerly had, the third partial has a frequency about a tenth of an octave below that which the fourth partial had, and so on. In the new sound, most partials are "replaced" by partials about a tenth of an octave lower in frequency, and the ear hears this as a fall in pitch. The fact that the weak first partial has disappeared, and that a new weak partial has appeared an octave above the old seventh partial, passes unnoticed.

Risset's "illusion" is closely related to an earlier illusion devised by the psychologist Roger Shepard, who used a computer to produce a succession of tones that seem to ascend endlessly in pitch by intervals of a semitone.

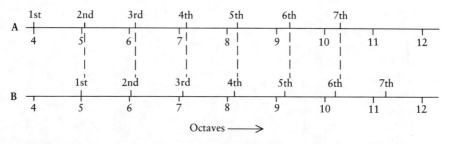

Octaves ⟶

Figure 14-5 Risset produced an interesting illusion in which, if all the frequencies present are doubled, the pitch goes down. The frequencies present are shown in part **A** (and downward dashed extensions), on an octave scale. The frequency of each partial is 1.1 octaves above that of the preceding partial. When all frequencies are doubled, as shown in part **B**, the listener hears each partial as being replaced by a partial a tenth of an octave lower in frequency, and so hears the pitch as going down. The ear doesn't notice that the original first partial has disappeared and that a new, high-frequency partial has appeared.

Shepard's illusion is explained in Figure 14-6. The envelope that specifies the intensity of a partial as a function of frequency is constant, as shown in parts A through E. In the shift from part A to part B, each partial goes up a semitone in frequency, so we hear a change in pitch of one semitone. However, the upper partials have become weaker, and after 12 such shifts, we arrive at the same configuration as shown in part A; so the pitch can continue to change without changing.

You need not proceed semitone by semitone, but can slide slowly up or down the scale, seemingly endlessly. Risset used this very effectively in incidental music for the play *Little Boy*, by Pierre Halet. The theme of the play is the recurrence of the Hiroshima bombing in a nightmare of Eatherly, the pilot of a reconnaissance plane. In this nightmare the bomb falls endlessly in a tone of ever-descending pitch.

Risset has also used tones in which the envelope shifts upward while the partials descend in frequency. The pitch decreases while the tone becomes shriller and shriller.

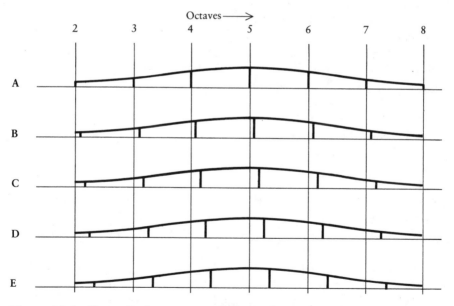

Figure 14-6 Shepard's famous eternally ascending tones. These tones are made up of octave partials whose amplitudes are specified by the fixed envelope shown in parts **A** through **E**. In the shift from part **A** to part **B**, each partial is increased in frequency by a semitone, and we hear a rise in pitch of a semitone. This recurs for the shift from part **B** to part **C**. After twelve increases by a semitone, we will be back at the same configuration as in part **A**; so, if we continue, we keep hearing a never-ending sequence of increases in pitch.

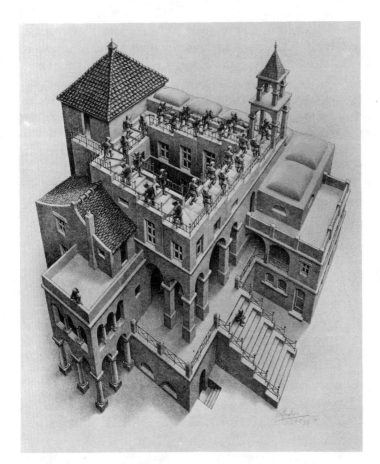

Figure 14-7 M. C. Escher's *Ascending and Descending.*

Kenneth Knowlton of Bell Laboratories and Risset have produced rhythmical sounds that constantly speed up but go slower and slower. As Figure 14-8 shows, the musical beat gets faster and faster, but the sixteenth notes gradually die out in intensity, to be replaced in turn by the eighth notes and the quarter notes.

Risset has produced subtle pitch paradoxes, in which the pitch of a bell-like sound with nonharmonic partials is easily identified with that of either of its two most intense sinusoidal partials.

We should note that the sounds described in Figures 14-5 and 14-6 differ from traditional musical tones, as do sine waves. In traditional musical tones, the underlying harmonic structure of Western music, the ratios of small integers (2:1, octave, 3:2, fifth, and so on) is present in the ratios of the frequencies of the harmonic partials that constitute the tone. In the sine wave itself there is just one frequency, and any frequency ratios

accelerando

Figure 14-8 By analogy with the process illustrated in Figure 14-6, Risset has produced a rhythm that goes faster and faster and yet slows down. Although the beat gets faster and faster, the shorter notes are gradually replaced by longer notes.

must be called up from memory. In the tone of Figure 14-5 there is a lowest partial, and an interval a little greater than an octave separates the partials. In the tones of Figure 14-6 there is no perceptible lowest partial, as there is in a musical tone. The octave is present, but no other ratios of small integers are represented in the partials.

It is plausible that a whole new music with new rules of consonance and dissonance could be developed around sine waves or the tones of Figures 14-5 or 14-6, but such a music would be very impoverished compared with traditional Western music. Rather, psychoacousticians use sine waves and tones such as those of Figures 14-5 and 14-6 to investigate human perception of unfamiliar stimuli and to try to sort out learned from inherent aspects of auditory perception. To these ends they also use the sequences of pulses and of tone bursts described in Chapter 7, for which the pulse or tone burst rate is not the fundamental frequency.

Illusions of Moving Sounds

Illusions of moving sounds are among the most effective of computer-produced effects. Much of the background has been covered in earlier chapters. Sound sources can be made to stand out, filling a room. Individual sources can whirl about over one's head as in John Chowning's *Turenas*. Such effects are most striking with four-track recording and four loudspeakers at the four corners of a square, as Chowning demonstrated. The position of the listener relative to the four loudspeakers may be less important for such "moving" sounds than for the reproduction of instrumental sounds by a quadraphonic system.

The principal effects to be produced are: (1) direction (azimuth) of the sound source; (2) distance of the sound source, and (3) motion of the sound source.

Chowning's prescription for direction is to divide the intensity of the sound source among two or four loudspeakers, using signals of the same phase or delay, but making the intensity greater from the loudspeakers in the direction the sound is to come from.

Chowning's prescription for distance is to control the ratio of reverberant to direct sound. We will hear more direct than reverberant sound from nearby sources; we will hear more reverberant sound than direct sound from distant sources. The overall amplitude of the direct sound should change as the reciprocal of the distance. The intensity should change as the reciprocal of the square of the distance. In a small room the intensity of the reverberant sound changes little with distance; in a large room it decreases somewhat with increasing distance of the source. Chowning has decreased the intensity of reverberant sound with the reciprocal of the distance of the source.

Slow source motions can be simulated simply by changing the direction and distance of the source. For swifter motion, the Doppler effect can be incorporated. The frequencies of sound sources moving toward us are raised; those of receding sources are lowered. A classical example is the sound of the whistle of a locomotive as it approaches, passes, and recedes. If the sound source approaches at a speed s, all frequencies are raised by a fraction s/v, in which v is the velocity of sound. If the source recedes at a speed s, all frequencies are lowered by a fraction s/v. To change the pitch by one semitone, the speed relative to the listener must be about 0.06 that of sound, that is, about 21 meters per second, or 68 feet per second.

In producing distance cues, some sort of artificial reverberation is necessary. In effect, fractions of the original computer-generated sound are added in with various delays. A great deal of study has been devoted to getting natural-sounding reverberation that does not alter the perceived spectrum of sounds. This can be done only if the reverberant signal has the same power spectrum as the original signal (colorless reverberation, first described by Manfred Schroeder) or if any alterations in spectrum change very rapidly with frequency, so that the reverberant intensity averaged over a critical bandwidth is little changed.

Although the best illusions of moving sound sources are attained with quadraphonic systems, startling effects are possible with only two loudspeakers. Not only do sound sources seem to move about; they stand out so that the sound fills the room in a way that seems unrelated to the positions of the loudspeakers.

Twice I have experienced the effect of sounds coming from unexpected directions. In the home of a friend I was listening to a recording of some rather ordinary music played on a synthesizer. The music was embellished with high-frequency chirping sounds. Some of these seemed to come from the left or right of the stereo loudspeakers, and even from the back of the room. On another occasion I was listening to a stereo radio broadcast that included high-pitched bird sounds. Some of these came from most unexpected directions, and certainly far to the right or left of the loud-speakers.

Perhaps in some sense, the sounds from the loudspeakers happened to mimic Schroeder's perfect stereo system (the one using a dummy head to pick up sounds, described in Chapter 12). Perhaps the explanation is simpler. The effect may have been caused by the directional patterns of the loudspeakers at very high frequencies, coupled with reflections from ceil-ing and walls. Perhaps both explanations are the same. What I *do* know is that a two-channel stereo system can cause high-frequency sounds to be heard as coming from almost any direction. I don't see why this effect could not be exploited using computer-generated sounds.

As noted early in Chapter 7, the pinnae of our ears are essential in judging the height of sound sources, and, presumably, in telling whether the source is in front of us or behind us. They can do this only by modifying the high-frequency spectra of the sounds that reach them; our judgment of up or down, ahead or behind, must be based on such modifi-cations. Surely it must be possible to modify the high-frequency spectrum of a sound deliberately, to give a sense of the sound's moving up, down, backward, or forward.

Categorical Perception

Whatever effects we may be able to attain by using computer-generated sounds, we want them to be of good quality. Early computer-generated sounds were buzzy or "electronic." Because our understanding of the nature of good instrumental sounds has been improved by computer analysis and synthesis of sounds, we now have computer-produced sounds that are neither harsh nor electronic. Some are almost indistinguishable from sounds of the instruments that they imitate. Some sound like nothing anyone has ever heard before, for they involve carefully controlled non-harmonic partials, or strange variations of partials that could not be pro-duced by conventional instruments or, in fact, any imaginable mechanical instruments. In some computer-produced sounds, one sound is trans-formed into another: a bell into prolonged liquid textures or into a group of singing voices, a voice into the roar of a lion. In others, "nonexistent" sound sources move through unoccupied space: ghostly instruments glide through an empty room.

As we have noted, some *sampling* digital systems can record and modify musical tones by transposition, imposition of a new envelope, extending in time, and other means. Or natural sounds can be combined, processed, and mingled with digitally generated sounds, to wonderful effect, as in Risset's *Sud*. In other cases, good musical sounds have been obtained by modifying speech or other sounds so profoundly as to completely disguise their origin.

People have learned to attain such effects through the exploration of sounds and through work toward characterizing such sounds. One fascinating property of sounds is *categorical perception.*

In everyday life we constantly assign objects to categories on a more or less correct basis. It is said that one of the earliest categorizations made by infants is *self-moving* (living) as opposed to *non-self-moving* (inanimate). *Animals* and *birds* are other simple categories.

The various distinct sounds of speech, called *phonemes*, form the "alphabet" of spoken language. Each language has a specific number of phonemes. The common consonant and vowel sounds of English are the phonemes of the English language. Phonemes differ from language to language.

The phoneme is the *percept*, not the physical sound. In different words the sound wave by which we recognize *b, g, k, o,* or *u* can differ. In common speech, the range of difference is small enough that we can recognize the phoneme correctly, whatever its context. We seldom confuse *got* with *cot*, even though the *g* (voiced) sound is very like the *k* (unvoiced) sound. Nor do we confuse *had* with *hod*.

When producing speech sound artificially, we can make a gradual transition between the sound wave characteristic of one phoneme and that characteristic of another. But when we do this, the listener hears *either* one phoneme *or* the other, not something in between. This is an example of categorical perception.

Categorical perception is merely categorical; it is not necessarily accurate. The hearer may hear a boundary-line sound sometimes as *g*, sometimes as *k*, but never something "in between." An English speaker has never learned anything in between, but only the limited number of phonemes characteristic of our language.

We are led to ask, Is there categorical perception of musical sounds? It is easy to jump to the conclusion that there is. An expert musician has no difficulty in saying, "That's a violin." Or a viola, or a French horn, or a saxophone. That sounds like categorical perception. But when John W. Grey synthesized musical tones midway between the cubes representing two instruments in the three-dimensional space shown in Figure 13-9, these tones weren't recognized as one instrument or the other. Rather, these sounds seemed related to the two instruments, or a mixture of them. This is contrary to the categorical perception of a speech sound as one phoneme *or* the other.

But musicians do treat pitch categorically. They will judge a tone of intermediate pitch as one note of the scale or the other, often not noticing whether it is flat or sharp. This is not a defect of hearing, for when they know what the task is, musicians with a good sense of pitch, on listening to a scale, can tell whether it is a just scale, a Pythagorean scale, or an equal-tempered scale.

Categorical perception occurs only when categories have been very firmly established, as in self-moving and non-self-moving, animals, birds, dogs, cats, the phonemes of a language, or the pitches of the diatonic scale. It is futile to seek categorical perception in distinct but unusual or miscellaneous sounds.

In conventional Western music, only certain pitches are intended (or "allowed"), just as in English only certain speech sounds are intended (or "allowed"). It therefore is not surprising to find categorical perception among pitches, and among chords as well. In Chapter 6, I noted that musicians correctly recognize the dominant seventh chord, even when it has been doctored into being acoustically consonant.

Categorical perception of timbres is another matter; yet I believe there is something of the sort, though it is weaker than with speech. We won't find this sort of categorical perception by wandering through the three-dimensional timbre space of Figure 13-9. That space is too limited. If there are musical categories of sound, they are not trumpetlike or trombonelike, but reedlike, brassy, bowed, struck or plucked, bell-like or gonglike (non-harmonic), drumlike, blocklike. We certainly recognize natural (noncomputer) sounds as belonging to various categories, which are associated with the nature of the sound-producing material and with its mode of excitation.

Categorical perception can be either a help or a hindrance. The scale and common chords give coherence to music. But it is tempting to think of *new* timbres that will *sound* new and different, and even of new scales and chords. We have learned language so thoroughly that we make only categorical distinctions between a relatively few speech sounds, hearing nothing in between. It is hard or impossible for an adult to undo this training. Adult Japanese find it very difficult to learn to hear the English *r* and *l* as different, though Japanese children learn readily. It may be difficult for us to hear, to distinguish, to recognize new and unfamiliar musical sounds. Happily, it seems not to be impossible.

Conclusion

In the nineteenth century, Helmholtz performed miracles — with very simple equipment — in the analysis and understanding of musical sounds. In the first half of this century, the electronic art derived from telephony made possible more quantitative and subtler experiments. In our own time the computer and other digital technology have made easy what was difficult or impossible with earlier electronic means. They have done something more: We can now transcend all the limitations of early sound sources. We can imitate the sounds of fine instruments. We can go beyond them. Our understanding of musical effects has increased, and so has our ability to produce them.

As I noted in the first chapter of this book, an increased capability to generate, experiment with, and understand sounds, new and old, has led some enterprising composers to pay more attention to the subtle qualities of the sounds used in their compositions. This seems to me to be a healthy alternative to excessive concern with formal structure, or to a search for "spontaneity" based on some form of improvisation.

Both rules and spontaneity have their place in music, but deep understanding and careful work are essential, too. However much we may wish that we could have heard the improvisations of Bach, Mozart, or Debussy, it seems likely that their best music is the music that they left us.

The genius of the past can be a weight upon the present. Perhaps by using new resources and new understanding, music can avoid being crushed by it. Scientists are not crushed by Newton and Einstein, for they have experimental resources, knowledge, and understanding that Newton and Einstein lacked. Whatever its "absolute" value, new science is new and worthy when it succeeds in going successfully beyond the old.

May it not be so with new music? But to succeed, new music must really be heard in the sense that its composers intended, must be understood, must hold the interest of and move the listener. Here an understanding and exploration of the science of musical sounds can help. The rest only talent or genius can supply.

Appendix A   Terminology

When scientists and engineers deal with well-defined, measurable physical properties, they use only unambiguous, well-defined terms, such as *time,* measured with a watch, *mass,* measured with a balance, or *length,* measured with a meterstick. This helps us grasp their intended meaning.

Many legitimate physical terms aren't nearly so simple. They can't be explained in a few words. Understanding them and using them properly results only from long exposure to experience or experiment, until the terms and their place in physics become familiar and, indeed, commonplace.

Efforts to "define" words briefly in terms of other words aren't very helpful. In both everyday life and science (if not in philosophy), we learn to use words understandably by protracted experience with things and by communicating with others. In this book I have tried to use words understandably, but it is difficult to eliminate every vestige of ambiguity.

In music we deal with many difficult qualities. They would not be made less difficult if I were to depart from the common words used by musicians, to invent a jargon, or to import one from psychology. I believe that the chief difficulties lie in the facts and experiences themselves, not in the words we use about them. I believe that the best "definitions" of the words that I use are in the text, explicitly or implicitly. Nonetheless, this brief discussion of terminology may prove useful to the reader.

Strictly, a *sound* is what we hear when a *sound wave* going through the air strikes our ears. A sound wave acts as what psychologists call a *stimulus*. Our *response* to the stimulus is the "sound" that we hear. By this definition, if no one is listening (or if only deaf persons are present), then there is no sound, but only a sound wave in the air.

The word *note* can designate *either* a mark on a musical staff *or* the sound produced, which we hear when someone "plays a note." Some try to avoid confusion by using the word *tone* for the sound produced when someone "plays a note."

A tone is a *musical* sound, one that can be heard as having a *pitch*. We can apply "tone" to the sound of a bell, but not to the sound of a drum. Musical sound waves are periodic fluctuations of air pressure. A *pure tone* is a sinusoidal sound wave. (One can also speak of the *good tone* of a violin, violinist, or pianist, but in this book I try not to use *tone* in that sense.)

Pitch is a quality that we hear in some sounds. Happily, for real periodic sounds the pitch that we hear is tied firmly to the periodicity, or frequency, of the sound wave. At concert pitch, A above middle C has a frequency of 440 Hz (vibrations per second). I therefore think it proper to designate pitch quantitatively by specifying frequency.

Loudness is how loud a sound sounds. It is related in a complicated way to the *intensity* of a sound. Intensity is measured in watts per square meter, a good, solid physical quantity.

223

Timbre is a quality that a sound has in addition to pitch and loudness. Sounds that don't have a clear pitch, such as those of drums and blocks, can differ in timbre. We can use many common words to distinguish timbres: *shrill, warm, harsh, dull, percussive.* Such words describe real, consistent differences in our responses to musical sounds and in the sound waves, but it is no simple matter to pin these differences down.

Is it differences in timbre that distinguish good from bad violin playing? Physicists rightly assert that a piano emits the same sound whether an expert player strikes a key or a weight falls on it; yet some pianists obviously have a "good tone" and others don't. I don't know how to explain this difference, and you won't necessarily get help by asking a pianist who has a "good tone." She (or he) can produce the effect, but will probably not be able to tell you in words how she does it. "Make it sing," Claude Shannon's clarinet teacher told him. Shannon knew what was wanted, if not how to do it.

Appendix B   Mathematical Notation

I have tried either to avoid mathematics in the text or to make it as simple as possible. Some mathematical expressions are necessary in conveying quantitative relations. For example,

   ml

means *m* times *l*. Clearly, this won't work for numbers, for we would not know whether

   27

was two times seven or twenty-seven. Hence, when we want to indicate the multiplication of numbers, we enclose them in parentheses. Thus

   (2)(7)

is 2 multiplied by 7. We can do this, if we wish, in multiplying quantities represented by letters,

   $(m)(l) = ml$

but nothing is gained by including the parentheses.

   The expression

   t^3

is the *third power* of *t* (or the *cube* of *t*). The meaning is

   $t^3 = ttt = (t)(t)(t)$

The *exponent*, 3, tells how many *t*'s to multiply together. A negative exponent indicates division rather than multiplication. Thus,

   $t^{-3} = 1/t^3 = (1/t)(1/t)(1/t)$

Let's consider a numerical example of powers. In an equal-tempered scale the frequency ratio of a semitone is

   $(2)^{1/12} = 1.059468$

Sometimes a very large number is written in the following way:

   $5.4 \times 10^5 = (5.4)(10)(10)(10)(10)(10)$
   $= 540,000$

We could write 5.4×10^5 as

   $(5.4)(10^5)$

but for some reason we don't. A very small number can likewise be written as

   $6.2 \times 10^{-4} = 6.2(1/10)(1/10)(1/10)(1/10)$
   $= 0.00062$

225

As a review,

$$ml/t^2$$

is the product of m and l divided by t^2.
The *square root* of a quantity, say, x, is such that

$$(\sqrt{x})(\sqrt{x}) = x$$

Thus,

$$\sqrt{x} = 2$$

and

$$(\sqrt{4})(\sqrt{4}) = (2)(2) = 4$$

We encounter decibels, abbreviated, dB, in several chapters. Decibels provide a way of expressing ratios of powers. If P_1 and P_2 are two powers (commonly measured in watts), P_2 is greater than P_1 by

$$10 \log_{10} (P_2/P_1) \text{ dB}$$

The logarithm to the base 10 of a number can be looked up in a table or obtained by using a "mathematical" hand calculator. Table B-1 gives the flavor of logarithms.

Table B-1 Logarithms and decibels.

Power ratio R	Amplitude ratio ($\sqrt{R}$)	$10 \log_{10} R$
.0001	.01	−40 dB
.001	.0316	−30 dB
.01	.1	−20 dB
.1	.316	−10 dB
1	1	0 dB
10	3.16	10 dB
100	10	20 dB
1,000	31.6	30 dB
10,000	100	40 dB
2	1.4	3 dB
1/2	.71	−3 dB

The MKS, or meter-kilogram-second, system of units is used in this book. The units in which quantities are measured are:

mass, m	kilogram
distance, l	meters
time, t	seconds
force, F	newtons
power, P	watts
intensity, I	watts per square meter
energy, E	joules

The pull (or "acceleration") of gravity at the earth's surface, which is 9.80 meters per second per second, produces a force of 9.80 newtons on a mass of one kilogram. Force is mass times acceleration.

A force of one newton pushing something a distance of one meter requires an expenditure of one joule of energy.

The number of watts is equal to the number of joules of energy expended per second.

In MKS units, pressure is measured in newtons per square meter. One newton per square meter is called a *pascal*.

Mathematics and Waves

Chapter 2 correctly describes the propagation of waves along a string or through the air as a traveling disturbance. Such a disturbance involves continual changes in momentum (mass times velocity) caused by a force. The force may be associated with the bending of a stretched string or the compression of air. Continual changes in the force occur because the stretched string is bent as the wave travels along, or because the air is compressed when the velocity associated with the wave is lower ahead than it is behind.

The propagation and properties of both transverse and longitudinal waves can be demonstrated by simple but rather tedious mathematical analysis (reasoning). One result of such mathematics is the discovery that waves behave in a very simple fashion only for small amplitudes; that is, when the stretched string along which a wave travels isn't bent too sharply or when a sound wave traveling through the air raises or lowers the pressure only by a small fraction. The behavior of such small-amplitude waves is *linear*. In essence, this means that when two waves are present in the same medium (string, air), they don't interact with one another. Each goes its own way as if the other weren't present. The total motion (displacement, velocity, or pressure) is simply the sum of the motions associated with the two (or more) waves.

We will not attempt a conventional mathematical analysis of waves here. Instead, we will assume that we are dealing only with small-amplitude linear waves. We will then ask, How can the velocity of the waves, the power carried by them, and other properties be expressed in terms of various physical properties? We do this by a seeming magic called *dimensional analysis.*

All physical quantities — including force, velocity, and momentum — have a dimension that is expressed in terms of the dimensions of time, mass, and length. Actual time measured in seconds is designated here by the italic letter t, whereas the dimension of time is designated by the boldface letter **t**, and similarly for mass, length, and any other physical quantity and its dimension. The three fundamental physical quantities and their dimensions are given in Table D-1. Let us illustrate the dimensions of some common physical quantities. What is the dimension f of frequency or periodicity in time? Frequency is the number of something per second. Number is dimensionless, and so the dimension of frequency is simply

$$\mathbf{f} = 1/\mathbf{t} \tag{D-1}$$

What about velocity? Velocity is distance per unit time. The dimension of distance is l, and the dimension of time is t, so p, the dimension of velocity, is given by

$$\mathbf{p} = \mathbf{l}/\mathbf{t} \tag{D-2}$$

228

Table D-1 Symbols for Physical Quantities and Their Dimensions

Physical Quantity	Symbol for Physical Quantity	Symbol for Dimension of Physical Quantity
Time	t	t
Mass	m	m
Length	l	l

Acceleration is change in velocity with time. Numerically, it is the amount that velocity changes in a unit of time. Dimensionally, the dimension a of acceleration is given by

$$d = v/t = l/t^2 \tag{D-3}$$

We now come to a physical law, not a matter of definition. This law was first stated by Newton. It is that, numerically, force is equal to mass times acceleration. Thus, the dimension F of force is

$$F = ml/t^2 \tag{D-4}$$

Energy (or work) can be defined as force times distance. Thus, the dimension of energy, E, is

$$E = ml^2/t^2 \tag{D-5}$$

Let us take a look at equation D-5. We note from equation D-2 that the dimension of velocity is l/t. Thus, the dimension of energy can also be written

$$E = mv^2 \tag{D-6}$$

Dimensionally, this is correct. But the numerically correct expression for kinetic energy, or the energy of mass in motion, is

$$E = (1/2)mv^2 \tag{D-7}$$

Here E is actual energy, not the dimension of energy, m is actual mass, and v is actual velocity. Dimensionally, equations D-6 and D-7 are in accord. The numerical factor $(1/2)$ is a number, and has no dimension. Finding the dimension of energy by dimensional analysis has led us to an expression for kinetic energy that is correct in everything except a multiplying numerical factor.

Let us keep what we have learned in mind, and see how we can use dimensional analysis in connection with waves.

As an example, what is the expression for the velocity v of a transverse wave traveling along a string of mass M kilograms per meter, a string that is stretched with a tension or force of T newtons?

First, what are the dimensions of M and T? M is mass per unit length; so the dimension of $\mathbf{M}$ is

$$\mathbf{M} = m/l \tag{D-8}$$

The tension T is simply a force, which has dimensions ml/t^2, so the dimension of $\mathbf{T}$ is

$$\mathbf{T} = ml/t^2 \tag{D-9}$$

The dimension of the velocity v of the wave must be l/t. We can see that this will be true if the expression for the velocity is

$$v = \sqrt{T/M} \tag{D-10}$$

We verify this by writing for the dimensions of $\sqrt{T/M}$,

$$\begin{aligned}
\sqrt{\mathbf{T/M}} &= \sqrt{(ml/t^2)/(m/l)} \\
&= \sqrt{l^2/t^2} \\
&= l/t = \mathbf{v}
\end{aligned} \tag{D-11}$$

Actually, equation D-10 is the numerically correct expression for the velocity of a wave traveling along a stretched string. The numerical factor turns out to be unity, but we couldn't know this from dimensional analysis.

Let us turn to something that is very important about plane waves traveling through air, that is, the intensity I, which is the power density measured in watts per square meter.

We noted earlier that energy or work can be defined as force times distance. Power is energy per unit time; so power will have the dimensions of force times (l/t), or force times velocity. Intensity I is power per square meter; so intensity will have the dimensions of force times velocity divided by l^2. Hence, the dimension $\mathbf{I}$ of intensity is given by

$$\mathbf{I} = \mathbf{F}(l/t)/l^2 = \mathbf{F}/lt = m/t^3 \tag{D-12}$$

But in plane waves in air we are concerned not with force F, but with force per square meter, or pressure, p. The dimension $\mathbf{p}$ of pressure is

$$\mathbf{p} = \mathbf{F}/l^2 \tag{D-13}$$

From these last two relations we see that the dimension $\mathbf{I}$ of intensity is

$$\mathbf{I} = \mathbf{p}(l/t) \tag{D-14}$$

Here $\mathbf{p}$ is the dimension of pressure, and l/t is the dimension of velocity. We may easily conclude that if p is the fluctuating pressure associated with a sound wave and u is the fluctuating velocity associated with a sound wave, the intensity of the wave will be

$$I = pu \tag{D-15}$$

This is not only dimensionally correct, but also numerically correct, and we could have arrived at it more directly.

In a small-amplitude linear sound wave, the pressure p will be some constant, which we will call K, times the velocity fluctuation u:

$$p = Ku \tag{D-16}$$

Thus, by using equation D-16, we can express intensity I in terms of either p or u:

$$I = Ku^2 \tag{D-17}$$

$$I = (1/K)p^2 \tag{D-18}$$

K is called the *characteristic impedance* or *wave impedance* for a plane sound wave. But how are we to find an expression for K? The dimension of K is

$$\mathbf{K} = 1/v^2 = l(l/t)^2 \tag{D-19}$$

Hence, from equations D-19 and D-12,

$$\mathbf{K} = m/l^2t = (m/l^3)(l/t) \tag{D-20}$$

The first factor on the far right of equation D-19 has the dimensions of mass density, which we will call D. The second factor has the dimensions of velocity. Can it be that

$$K = Dv \tag{D-21}$$

in which D is the density of the air and v is the velocity of sound? It can be and it is, equation D-21 is numerically correct. This is very plausible, for from equations D-21 and D-17 we can write

$$I = (Du^2)v \tag{D-22}$$

Du^2 is proportional to the kinetic energy per cubic meter of the air moving at a velocity u, and in some sense this energy is transported through the air at a rate v. The kinetic energy is only half the energy transported; there is an equal amount of potential energy associated with the compression of the air by the sound wave.

We can also express the intensity of the sound wave in terms of the pressure p as

$$I = p^2/Dv \tag{D-23}$$

For air at 20°C,

$$D = 1.2174 \text{ kilograms per square meter}$$

$$v = 344 \text{ meters per second} \tag{D-24}$$

$$I = 0.002388 \, p^2$$

We should note that for a fluctuating pressure, the average intensity is given by the average value of p^2 divided by Dv. For a sinusoidal variation of pressure with time, the average value of p^2 is half the square of the peak pressure (That is, the "top" of the sine wave).

In Chapter 7 the reference level of intensity is given as 10^{-12} watt per square meter. The usual reference level is a pressure of 0.00002 pascal (a pascal is a pressure of one newton per square meter). If we calculate I for this pressure using (D24), we get

$$I = 0.955 \times 10^{-12} \text{ watt per square meter}$$

This is so close to the 10^{-12} watt per square meter (only 0.2 dB different) that I chose to use the round number 10^{-12} watt per square meter in Chapter 8.

In Chapter 8 I commented on the sensitivity of the ear, and said that "theoretically" we should just be able to hear a 1-watt, 3,500-Hz sound source at a distance of 564 kilometers (352 miles). If a sound has a power of W watts and travels out equally in all directions, so that at a distance L the power passes evenly through a sphere of area $4\pi L^2$, the intensity I at a distance L must be

$$I = W/4\pi L^2 \tag{D-25}$$

If we let $W = 1$ and $L = 564,000$ meters, we get from equation D-22 just about 10^{-12} watt per square meter, which is about the threshold of hearing.

Let us turn to the velocity with which a sound wave travels through air. This velocity does not vary with pressure, but it does vary with temperature. The pressure of air is caused by the velocity of air molecules. The square of this velocity is proportional to the temperature in kelvins, that is, in degrees measured with respect to absolute zero; that is, 0 kelvin is −273 degrees Celsius (centigrade).

It is dimensionally plausible that the velocity of a sound wave should be proportional to the velocity of the molecules of the air through which the wave travels, and this turns out to be so. The velocity of a sound wave at a temperature T can thus be expressed as

$$v = v_k \sqrt{T/T_k} \tag{D-26}$$

in which v_k is the velocity of the sound wave at a temperature of T_k kelvins. If we take the velocity of the sound wave to be 344 meters per second at a temperature of 20 degrees Celsius (centigrade), then at a temperature T kelvins,

$$v = 344 \sqrt{T/293}$$

$$v = 20.1 \sqrt{T} \tag{D-27}$$

The velocity of sound varies with humidity as well as with temperature. The velocity of the molecules of a gas varies with the mass of the molecules as well as with the temperature: the lighter the molecules, the greater their velocity. Molecules of water vapor are less massive than molecules of dry air, and so the velocity of sound increases with increasing humidity.

Wind instruments are provided with tuning adjustments to compensate for the effects of temperature and humidity on pitch. The pitch of a pipe organ inexorably changes with temperature and humidity, and other instruments must accommodate this.

Appendix E Reflection of Waves

In order to understand the reflection of sound waves, we must take into account the fact that such waves consist both of an increase or decrease in the pressure of the air, designated by p, and of a forward or backward velocity of the air, designated by u. As noted in Appendix D, these two components of the wave go hand in hand. In a wave that travels from left to right,

$$p = Ku$$

in which K is a constant.

Figure E-1 shows graphically the pressure p and the velocity u of a "square" wave traveling to the right. Part A depicts a wave in which the pressure p is positive. The associated velocity u is positive; that is, it represents motion of the air to the right, in the direction in which the

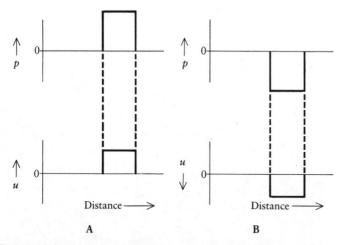

Figure E-1 For a "square" sound wave traveling to the right, the pressure p and the velocity of the air u are plotted against distance. When the pressure is greater than average, the pressure difference p is positive and lies above the axis, the horizontal line marked 0. When u lies above the axis, the velocity is to the right; when u is negative (plotted below the axis), the velocity is to the left. For a wave traveling to the right, the pressure p and the velocity u have the same sign; that is, if one lies above the axis, both do. When p is positive, u is positive, as shown in part **A**. When p is negative, u is negative, as shown in part **B**.

234

wave travels. In part B, the pressure p is negative, and is represented as lying below the axis, the horizontal line marked 0. This merely means that the total pressure of the air is less than the average air pressure. The velocity u is also negative (motion of the air to the left), and this is shown as a velocity lying below the axis.

   Figure E-2 shows p and u for a square wave traveling to the left. Why do the graphs of waves traveling to the right look so different from those of waves traveling to the left? A wave can travel to either right or left. Shouldn't the picture be much the same in either case? The difference arises because in both cases we measure a velocity as positive (above the horizontal axis) if it is directed toward the right, regardless of whether the wave actually travels to the left or to the right. In any sound wave, if the pressure p is positive, the velocity of the air, u is in the direction in which the wave travels; if the pressure p is negative, the velocity of the air is directed opposite to the direction of travel of the wave. Thus, in the graphs of a wave traveling to the left, the pressure p is positive and the velocity u is negative, that is, to the left, the direction in which the wave travels.

   Graphs such as these are very useful for understanding the reflection of waves. Figure E-3 illustrates successive stages in the reflection of a wave

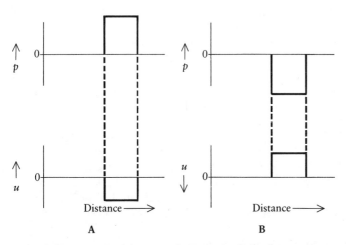

Figure E-2   For a wave traveling to the left, the pressure p and the velocity u have opposite signs. When the pressure is positive (as in part **A**), the velocity is negative. When the pressure is negative (as in part **B**), the velocity is positive. This is so because a positive velocity is generally depicted as a velocity to the right. Thus in part **A** the pressure is positive, and the velocity is in the same direction as that in which the wave travels, that is, to the left, shown as a negative velocity.

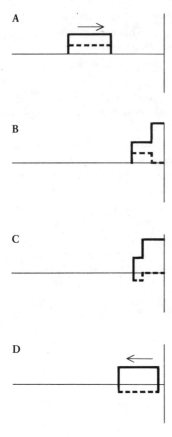

Figure E-3 The reflection of a sound wave from a solid wall, represented by the vertical line to the right. Pressure is represented by a solid line, and velocity by a dashed line. Part **A** represents a wave traveling to the right, toward the wall; part **D**, a wave traveling to the left, away from the wall; parts **B** and **C**, the wave in the process of reflection. The complicated curves of pressure and velocity in parts **B** and **C** are simply combinations of the pressures and velocities of waves traveling to the right (the *incident* wave) and to the left (the *reflected* wave). The velocities of these two waves must be such that at the wall the sum of the velocities is zero, for the air can't move at the wall. This simple condition determines the velocity, and hence the pressure, of the reflected wave.

from a wall, represented by the vertical line at the right. In this figure, pressure p is represented by a solid line and velocity u by a dashed line. In part A, we see the pressure and velocity of an *incident* wave approaching from the left. Both are positive, as is proper for a wave traveling to the right. In part D, the wave has been *reflected* from the wall. The pressure is positive and the velocity is negative, as is proper for a wave traveling to the left.

What of the pressure and velocity during the process of reflection, as shown in parts B and C? Here the pressure near the wall is twice as great as that of the incoming wave, and the velocity near the wall is zero. Can we explain this complex behavior?

The behavior is actually very simple. During the process of reflection we have overlapping waves, one traveling to the right and the other to the left. The pressure and velocities shown in parts C and D are the sums of the pressures and velocities of these two waves: The pressures add, but the velocities, being in opposite directions and therefore having opposite signs, cancel each other out.

We see that, in reflection from a solid obstacle, the pressure of a sound wave is reversed. When a sound wave travels through a tube, we can have another sort of reflection. If the end of the tube is open, there can be no (or very little) pressure at the end of the tube, but we can have a velocity u at the end of the tube.

Imagine a sound wave traveling through a narrow tube, such as an organ pipe. When it reaches an open end, it will be almost completely reflected. But, after reflection at an open end, the air velocity u will be unchanged and the sign of the pressure p will be reversed. If the pressure p is positive before reflection, it will be negative after reflection.

In relating the length of a pipe to pitch, we must take into account the nature of the reflections at its ends. Figure E-4 illustrates successive reflections in an organ pipe open at both ends (left) and in a pipe open at one end and closed at the other (right).

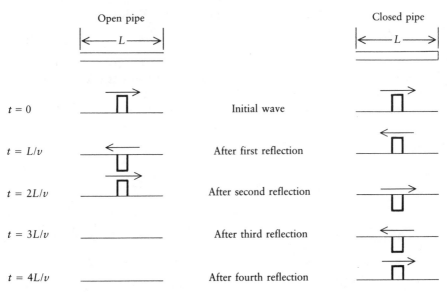

Figure E-4 The pressure of a wave after successive reflections at the ends of an open organ pipe (open at both ends, left) and a closed organ pipe (open at one end, closed at the other, right). In the open pipe, after two reflections the pressure is back where it started, and the pitch frequency is $v/2L$ Hz. In the closed pipe, the pressure is back where it started only after four reflections, and the pitch frequency is $v/4L$ Hz. For the same pitch, open organ pipes must be twice as long as a closed pipe.

In reflection from an open end, the pressure p changes sign on reflection, but, in reflection from a closed end, the sign of the pressure p remains the same after reflection. For an open pipe, we see that after a time,

$$2L/v$$

the wave has undergone two reflections and is the same as at the start. However, for a closed pipe the wave must undergo *four* reflections before it is the same as at the beginning, and this takes time.

$$4L/v$$

Thus the pitch frequency f of an open organ pipe is

$$f = v/2L \text{ Hz} \tag{E-1}$$

whereas the pitch frequency f of a pipe closed at one end and open at the other is

$$f = v/4L \text{ Hz} \tag{E-2}$$

Here v is the velocity of sound and L is the length of the pipe.

In stringed musical instruments, both ends of the string are rendered immovable. Hence, at the ends of the string, the transverse velocity is always zero. All reflections are the same, and the pitch frequency is always

$$f = v/2L \text{ Hz} \tag{E-3}$$

But, in this case, v is the velocity with which a transverse wave travels along the string. As noted in Appendix D, this velocity is higher the greater the tension, and lower the greater the mass of the string.

Appendix F Digital Generation of Sound

Initially, the digital generation of musical sound was accomplished on general-purpose computers through special software. The generation of one second of sound commonly took several tens of seconds of computer time.

Today a host of digital synthesizers that run in real time are available. In their organization and function these draw on techniques first developed in software, such as Max Mathews's Music V, described in more detail in his 1969 book, *The Technology of Computer Music.* Music V and other software programs derived from it are still used in digital sound synthesis and in the processing of natural sounds. It seems appropriate to use Music V in explaining the capabilities and problems of digital synthesis.

How is it possible for a computer to generate sounds? The *sampling theorem* gives us a clue. Consider any waveform made up of frequency components whose frequencies are less than B. That is, consider any sound wave whose frequency components lie in the bandwidth between zero and B. *Any* such waveform or sound wave can be represented *exactly* by the amplitudes of $2B$ *samples* per second. These samples are merely the amplitudes of the waveform at sampling times spaced $1/2B$ apart in time. Part A of Figure F-2 represents a waveform. In part B, the amplitudes at times $1/2B$ are shown as vertical bars drawn from the horizontal axis (which represents zero amplitude) to the curve of these bars. The successive heights can be represented by $2B$ *numbers* each second. These numbers describe the required samples from which the waveform can be reconstructed. Part C is an exact replica of the waveform in part A. This replica can be obtained by passing short electric pulses of the heights given in part B and described by the numbers that express these heights, through a low-pass filter of bandwidth B.

In high-quality computer-generated sound, it is customary to use 44,100 samples per second to represent a waveform. This allows us to produce frequencies up to 22,050 Hz. Because of technological limitations, the frequency range of bandwidth actually attained is 20,000 Hz.

The samples in part B of Figure F-1 are shown as lines of various heights, and the succession of sample heights stands for a succession of numbers. A computer can't produce every exact number, for most numbers can be represented only by an infinite number of digits to the right of the decimal point. A computer can produce a set of numbers such as 00 (0), 01 (1), 05 (5), 27, 44, 99 — all of which are examples of the 100 possible two-digit numbers starting with 00 and ending with 99.

The internal organization of computers is such that they use only two *binary digits*, 0 and 1, and represent numbers in terms of them.

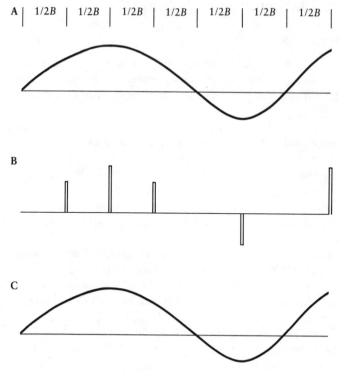

Figure F-1 A waveform of bandwidth B can be represented *exactly* by $2B$ samples a second, taken at time intervals $(1/2B)$. In part **A**, the samples are represented by the lengths of vertical lines drawn from the horizontal line to the curve representing the waveform. The sample amplitudes may be described by numbers. In part **B**, the sample amplitudes are represented by very short electric pulses whose heights correspond to the sample amplitudes. In part **C**, the original waveform is recovered by low-pass filtering of the sequence of pulses shown in part **B**.

The successive digits of common or *decimal* numbers are interpreted by means of powers of 10. Thus, 257 means

$$7 \times 10^0 + 5 \times 10^1 + 7 \times 10^2$$

$$= 7 \times 1 + 5 \times 10 + 7 \times 100$$

The binary number 1001 means

$$1 \times 2^0 + 0 \times 2^1 + 0 \times 2^2 + 1 \times 2^3$$

$$= 1 \times 1 + 0 \times 2 + 0 \times 4 + 1 \times 8$$

In decimal notation, the binary number 1001 is 9.

In good computer-produced sounds, 16 binary digits are used to represent the samples. This enables us to represent 65,536 different sample amplitudes. In effect, half of these binary numbers are used to represent positive sample amplitudes, and half are used to represent negative sample amplitudes.

If we use 16 binary digits to represent the largest possible sine wave, the signal-to-noise ratio of the sine wave so produced will be about 98 dB.

Suppose that we choose 44,100 sample amplitudes a second. We then produce a sequence of 44,100 short electric pulses a second, such that the amplitude of each is equal to the sample amplitude. We then pass this train of pulses through a low-pass filter to eliminate any frequency components above 22,050 Hz. Because we are free to choose the numbers that give the sample amplitudes *in any way that we wish,* we can by this means produce *any possible* sound wave whose bandwidth is 22,050 Hz or less, at least with the accuracy indicated. This is good enough for high-quality musical sounds.

This is too much freedom of choice to be of any use. In Music V, Mathews found a way to program a computer to produce a wide variety of musically useful sounds. Indeed, subsequent experience seems to indicate that Music V can be used to produce sounds as elaborate as we wish, including the sound of a singing voice and even speech sounds.

How is this done in Music V? We have all heard of the idea of a computer *simulating* some sort of mechanical or electrical device. The easiest way for noncomputer people to understand Music V is to think of it as making the computer simulate the operation of several fundamental electronic devices that are connected together in various ways. These fundamental devices are shown in Figure F-2.

One important device is the *oscillator*, represented in part A. This has an output and two inputs. The number going into input I1 specifies the amplitude of the output wave; the number going into I2 specifies the frequency of the output wave. Each oscillator is programmed to produce a particular waveform *Fn*, which may be a sine wave, a square wave, or some other wave.

The *adder* shown in part B is essential. The output of the adder is the sum of the two inputs, I1 and I2. We may use the adder to add the outputs of two sinusoidal oscillators to get the sum of two partials. Or we may use the adder to add a small sinusoidal vibrato to the number that specifies the average frequency of the oscillator.

The output of the *multiplier* shown in part C is the product of the two input numbers I1 and I2. The multiplier may be used in several ways: to multiply by some chosen number the amplitudes of all outputs produced by an oscillator, so that we can use one number as a volume control; or to change the frequency of an oscillator by a constant factor, thus transposing any notes played.

The final device (part D) is an output device that stores the sequence of numbers that represent the samples of the waveform produced. This storage may be accomplished in the computer memory, on disk, or on tape.

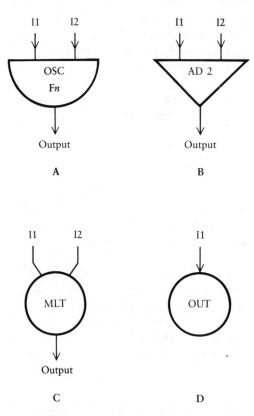

I1   I2

OSC
F*n*

Output

A

I1   I2

AD 2

Output

B

I1   I2

MLT

Output

C

I1

OUT

D

Figure F-2 The four devices that Music V "simulates." Part **A** represents an oscillator *Fn* whose frequency is controlled by input number I2 and whose output amplitude is controlled by input number I1; part **B**, an adder whose output number is the sum of the two input numbers I1 and I2; part **C**, a multiplier whose output number is the sum of the input numbers I1 and I2; and part **D**, an output device to store a sequence of numbers that represent sample amplitudes.

Figure F-3 shows how to program an *instrument* and play two notes. This instrument makes use of two oscillators. Oscillator F2 produces the "squarish" waveform shown in part C. Oscillator F1 produces a single time-varying output that rises from zero and falls to zero again, as shown in part B. This output determines how the amplitude of the output of oscillator F2 rises and falls with time. Thus, the amplitude and duration of the note produced by F2 are controlled by the input numbers P5 and P6. The frequency of the note is controlled by input P7.

The program for "creating" the instrument and for playing the two notes shown in part D is given in lines 1 through 10 of the figure. In this program, lines 1 through 5 define the instrument shown in part A. Line 6

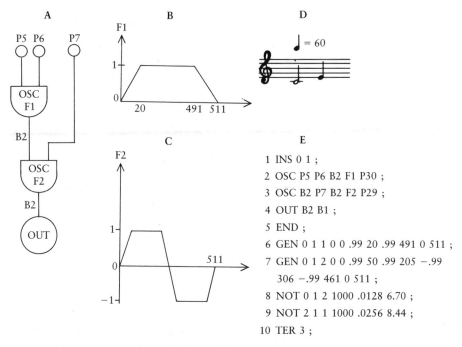

A   B   D

F1

P5 P6   P7   ♩ = 60

OSC
F1

1
0
20   491 511

B2   C   E

F2   1 INS 0 1 ;

OSC   2 OSC P5 P6 B2 F1 P30 ;
F2   3 OSC B2 P7 B2 F2 P29 ;

B2   4 OUT B2 B1 ;
                                   5 END ;
OUT   6 GEN 0 1 1 0 0 .99 20 .99 491 0 511 ;
                          1   7 GEN 0 1 2 0 0 .99 50 .99 205 −.99
                          0   511   306 −.99 461 0 511 ;
                          −1   8 NOT 0 1 2 1000 .0128 6.70 ;
                                   9 NOT 2 1 1 1000 .0256 8.44 ;
                                   10 TER 3 ;

Figure F-3   A simple instrument, and the program that causes it to play the
two notes shown in part **D**. The instrument consists of an oscillator, F1, that
produces the envelope (part **B**) of the output of oscillator F2. The waveform
of oscillator F2 is shown in part **C**. Lines 1 through 5 describe the instrument;
lines 6 and 7 describe the envelope and waveform; and lines 8 through 10 play
the notes.

defines the time function (part B) that oscillator F1 produces. Line 7
defines the waveform (part C) that oscillator F2 produces. Lines 8 and 9
cause the instrument to play the two notes shown in part D.

In line 8, the 0 following NOT says to start this note at a time zero.
The 1 that follows says that the note is to be played by instrument
number 1 (defined in lines 1 through 4). The 1000 that follows specifies
the amplitude of the output. The following number, .0128, is, in fact, the
input number P6 of part A. When this number times the number of
successive samples is equal to 511, the output of oscillator F1 will have
risen from zero and fallen to zero again, as shown in part B and line 6.
The first note is two seconds long, and it is assumed in the example that
there are 20,000 samples per second. The entry for P6 in line 8 is .0128.
We note that

$$(.0128)(2)(20{,}000) = 512$$

which is close enough.

The last number in line 8 is the input P7, and this specifies frequency. According to part C and line 7, one complete cycle happens when P7 times the number of samples equals 511. Hence, the number of cycles per second, or Hz, is P7 times 20,000, so that the frequency is

$$(6.70)(20,000)/(511) = 262 \text{ Hz}$$

This is indeed the frequency of middle C.

The NOT lines of the figure are a very primitive way of playing a very primitive instrument. The next-to-last number, P6, depends on note duration. Surely the computer can compute this for us! Indeed, it can and does. If we wish, we can enter the frequency instead of P7, and the computer can compute the required P7. Or we can enter the name of the note and its octave number, or the octave number and a number of semitones. We can have much more complicated instruments that make it possible to have vibrato, to sweep the frequency smoothly with time, to do a host of things.

Let us just take it for granted that we can do anything we want with Music V. What do we *want* to do?

There are several methods of sound synthesis. The most direct is to add together a lot of sinusoidal partials whose amplitudes rise and fall somewhat differently for the duration of the note. This is called *additive synthesis*. It is very powerful but somewhat slow, because each sinusoidal partial must be computed separately.

In the early days of sound synthesis, the oscillators of Music V were programmed to produce geometrically simple waveforms, such as those shown in Figure F-4. This was economical, but the sounds produced were limited and inflexible in quality, and not very good.

In his book *The Technology of Computer Music*, Mathews describes an instrument whose waveform varies with amplitude (see Figure F-5). Such an instrument has features that ordinary musical instruments have: The quality of the tone changes as the intensity of the tone is increased; louder tones have more partials than weaker tones. Figure F-6 illustrates what can be accomplished with an instrument of this sort. Curve A is a plot of the nonlinear relation between the output amplitude and the input

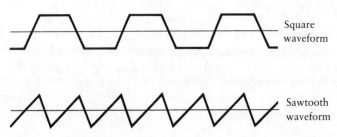

Square waveform

Sawtooth waveform

Figure F-4 Two simple waveforms that can be produced by using Music V.

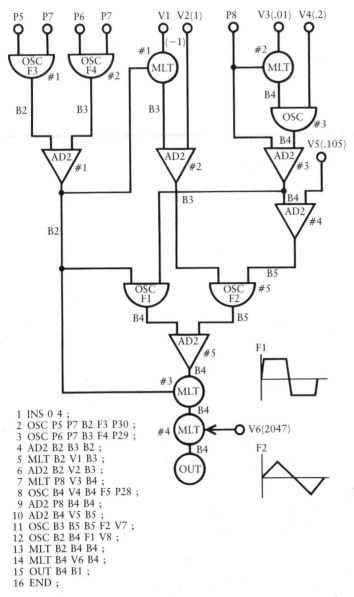

```
1  INS 0 4 ;
2  OSC P5 P7 B2 F3 P30 ;
3  OSC P6 P7 B3 F4 P29 ;
4  AD2 B2 B3 B2 ;
5  MLT B2 V1 B3 ;
6  AD2 B2 V2 B3 ;
7  MLT P8 V3 B4 ;
8  OSC B4 V4 B4 F5 P28 ;
9  AD2 P8 B4 B4 ;
10 AD2 B4 V5 B5 ;
11 OSC B3 B5 B5 F2 V7 ;
12 OSC B2 B4 F1 V8 ;
13 MLT B2 B4 B4 ;
14 MLT B4 V6 B4 ;
15 OUT B4 B1 ;
16 END ;
```

Figure F-5 A more complicated instrument, whose waveform depends on the amplitude, and on the program that defines it.

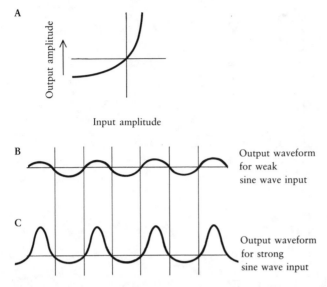

Figure F-6 A waveform that changes with amplitude can be produced by using a sine wave as an input to a device with the input-output characteristic shown in part **A**. If the input is a sine wave of small amplitude, the output will be almost sinusoidal, as shown in part **B**. If the input is a sine wave of larger amplitude, the output will be peaked waveform, as shown in part **C**; this has many harmonic partials.

amplitude. Waveform B shows the output waveform when the input is a weak sine wave; the output waveform is almost sinusoidal and has chiefly one partial. As shown in part C, the output for a strong sinusoidal input is peaked, and has many harmonic sinusoidal partials. The stronger the input, the stronger the higher partials. Composers Marc Lebrun and Daniel Lafitte have used this means to obtain rich tones whose timbres change with amplitude.

John Chowning invented another method, which easily produces rich, brassy sounds, called *fm* (frequency modulation) *synthesis* (see Chapter 13). It is done, in effect, by putting a sinusoidal vibrato on a sinusoidal oscillator, but the vibrato has the *same frequency as the oscillator*. In part A of Figure F-7, the amplitude of the output is plotted against a sort of pseudotime t', which governs the rate of generation of samples. Curve B shows the effect of the vibrato on the rate of change of t' with actual time t. At points p, t' changes rapidly with t; at points v, t' changes slowly with t. The resulting output waveform is shown in part C. The peaks rise and fall rapidly; the valleys fall and rise slowly. The resulting waveform has many high-frequency harmonics. Moreover, the intensities of these high-frequency partials can be increased by increasing the amount of frequency

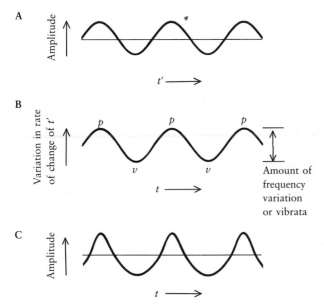

Figure F-7 Chowning's fm synthesis. In part **A** is a sine wave whose amplitude is a function of a variable t'. Part **B** shows the rate at which t' changes with time, t. This rate has a sinusoidal variation with time, a sort of vibrato. In part **C**, the waveform of part **A** is plotted, not against t', but against actual time t. The waveform is peaked and has many harmonic partials. By gradually increasing the amount by which t' varies with time, we can gradually increase the strengths of the higher partials.

modulation (vibrato). If we gradually increase the amount of frequency modulation during the playing of a note, the intensities of the higher partials gradually increase. This is characteristic of trumpet tones. Of course, the amplitude of the sine wave that is frequency modulated in this way should rise and fall during the playing of the note.

Other interesting effects can be obtained by frequency modulating with a frequency that is an integer multiple of the oscillator frequency, or with a frequency that is not quite equal to the frequency of the oscillator.

In the early days of computer music, a few centers exchanged software and modified it. Music V and its derivatives offered a standard of sorts, but actual programs were different for different machines. So were formats for inputting data, and means for converting digits into sound.

These software synthesis programs enabled the user to generate new instruments with new sound qualities and new control parameters. A strong emphasis on sound itself was built into such programs. In contrast, MIDI (Music Instrument Digital Interface) is fundamentally keyboard oriented. It provides a way to play notes, and it provides ways to operate whatever controls a synthesizer may have.

As various keyboard synthesizers began to appear, users wanted them to be connected and communicate with each other. There was strong motivation for an industry-wide standard.

MIDI was the response to this need, and it is the most important standard in computer music today. It came about through initial discussions among several synthesizer manufacturers in 1981. In 1983 an official description, called the MIDI 1.0 "Detailed Specification" was released.

The design of MIDI reflects its early concept as a keyboard standard. But MIDI proved to be adaptable to many functions, to the control of a host of synthesizer functions, and to the control of synthesizers by means other than keyboard: by simulated wind and string instruments, by sequencers or computers, or by devices such as Max Mathews's Radio Baton.

In essence, MIDI is both a hardware standard and a protocol for communication between devices, whether they are synthesizers, computers, sequencers, drum machines, or any device that is designed to receive and/or respond to MIDI commands.

MIDI commands can be used to turn a tone on (a separate command is needed to turn the tone off), to change the tuning or timbre, and for many other purposes, some of them *system exclusive*, that is, pertaining to the equipment of a particular manufacturer.

The MIDI physical connection is simple and inexpensive. Connectors on devices are standardized; they are five-pin DIN female connectors, with pins 1 and 3 not used. Pins 4 and 5 transmit the MIDI signal. Pin 2 connects to a wire in the cable that is wrapped around the others and shields them; it is connected to ground.

Nearly all synthesizers and studio gear have MIDI connections built in. There are three possible connections: *MIDI In, MIDI Out*, and *MIDI Thru*, the last providing an exact copy of what *MIDI In* receives. By means of MIDI Thru a number of devices can be connected together in a daisy-chain fashion. Or you can use a splitter that provides several *MIDI Outs* from one *MIDI In*.

It is not so easy to control a MIDI device from more than one MIDI source—for example, to control a synthesizer from both a sequencer and a keyboard. A special "merge box" can be used that will preserve the integrity of each MIDI source. Many computer-based sequencers and many control devices have a merge function that will combine MIDI information from sequencer and keyboard so that a merge box is not needed.

Computers may or may not have MIDI connections built in, but the MIDI interfaces built for personal computers are quite inexpensive. They plug into the serial port(s) of the computer.

The MIDI signal is bit serial, a 5-milliampere current loop, opto-isolated, running at 31.25 kBaud (31.25 thousand bits a second). It is asynchronous, that is, a byte or burst of digits can be sent at any time, but always at the 31.25 kBaud rate.

Given the hardware, let us examine the protocol for MIDI commands.

Each MIDI command consists of a *Start* bit (to alert the receiver), eight data bits, and a *Stop* bit, a total of ten bits per serial byte; so the transmission time per byte is 0.00032 seconds. The device to be controlled discards the *Start* and *Stop* bits and interprets the remaining eight-bit message byte. There are two kinds of bytes:

Status Bytes—any byte starting with a 1

Data Bytes—any byte starting with a 0

This leaves seven bits per byte for *Status* or *Data*.

A MIDI message or command usually consists of a *Status Byte*, to tell what function is addressed, for example, *Program Change* (choose a timbre. "Big Brass," "Heavenly Choir," and so on), *Note On* (start playing a note) or *Note Off* (stop playing a note). Such *Status Bytes* will be followed by zero or more *Data Bytes* telling, for example, which timbre to choose, or what note to play and how loudly (with what *Key Velocity*) to play it. A *Note On* followed by data saying that the key velocity is zero is equivalent to a *Note Off*.

Control Change can be used to add nuance to a sound, as vibrato, tremolo, sustain pedal, soft pedal.

After Touch provides a way to control the sound after it has been initiated.

Pitch Bend will alter the pitches of all sounding notes on a given channel.

In early keyboard synthesizers, all notes were played with the same timbre. In good present-day synthesis, one synthesizer can be configured to respond to several different MIDI channels simultaneously, with different timbres and different expressive effects on different channels. MIDI provides for sixteen control channels. A given synthesizer can be set to operate in one of four *Channel Modes*:

MODE 1: Omni On, Poly. Receive on all channels, can play more than one note at a time. "Goof proof" for beginners.

MODE 2: Omni On, Mono. Receive on all channels, but play only one note at a time, as with the older monophonic synthesizers.

MODE 3: Omni Off, Poly. Receive only on specified MIDI channels, each of which can play a different timbre and more than one note at a time. Typical for use with sequencers. This has become the most common mode.

MODE 4: Omni Off, Mono. Receive only on specific channel. Typically, you would use this mode with a guitar controller; you would have one MIDI channel for each string of the guitar. This gives the ability to add individual expression in each channel.

When the *Mode* has been chosen, you must still play notes. *Channel Voice Messages* consist of a status byte followed by one or more data bytes. They affect only those (*Omni Off*) synthesizers that are set to receive on specified channel(s) by a *Channel Mode* message. *Channel Voice* messages are basically everything you play on the keyboard, for example, which key is pressed and which key is released, which are *Note On* and *Note Off*, respectively. Some *Channel Voice Messages* encode the note number as well as channel, and some are "channel-wide" (these do not encode note number).

System Messages (not encoded with channel number) can be received regardless of MIDI *Channel Mode* set. These commands control the whole MIDI system, for example, to synchronizing a sequencer and a drum machine, start and stop performance, and other uses.

These are some examples of *System Message* commands:

System Common — Status byte followed by zero or more data bytes. Examples are: *MIDI Song Position Pointer* (location in a sequence) and *MIDI Song Selection* (sequence number).

System Real Time — A single status byte. These commands are used to start, stop, continue, and synchronize sequencers or drum machines.

System Exclusive — A status byte, followed by *Manufacturer's ID*, followed by any number of data bytes, followed by *EOX* (End of Exclusive status) byte.

System Exclusive commands are specific to each machine. The "trapdoor" of MIDI, *System Exclusive* commands let you do "bulk dumps" of synthesizer parameters, or even of samples of digital audio. (It is not immediately obvious that MIDI can be used to transfer sounds!)

Several important additions have been made to the original MIDI standard. First, the *Sample Dump Standard (SDS)*, adopted in 1986, allows a more convenient way to upload or download sound files from computers. This is a non-real-time process, but still useful for creating, preparing, and editing sounds on a host computer to be used in a sampler.

Another supplement to MIDI is *MIDI Time Code* (MTC). This adds an *absolute* time reference to the standard. The basic idea is to encode *SMPTE* code (Society of Motion Picture and Television Engineers) synchronization standard) over MIDI. Synchronization has become extremely important in video post-production now that using a sequencer or computer to control synthesizers by MIDI has become a major activity.

It is important to realize that MIDI commands have no sense of time at all. When a MIDI command goes over a wire, it is executed as soon as it arrives at its destination. There is no intrinsic way to "save" a MIDI message, to be executed later. Saving MIDI messages is done by sequencers or computers that form a part of the system and schedule the MIDI events.

Although the abbreviation *MIDI* implies some form of digital music, it is significant that no mention is made in the specifications of what will actually *make* the sound. It could be a MIDI-enabled acoustic piano, or a MIDI-controlled carillon, calliope, drum set, merry-go-round, or rocket launcher. That is to say, the sound-producing apparatus can just as easily be mechanical as electronic. A near-ideal application of MIDI is the "MIDI piano" that has appeared recently, such as the *Diskclavier* from Yamaha. In this context, it appears that it is possible both to capture and regenerate (or analyze) full concert performance via MIDI. Of course, this is really a modern version of the piano roll, but the information is easily edited, copied, and stored as MIDI data in a personal computer.

Phenomenally successful, MIDI has been built into the whole digital sound industry. Yet it has its limitations, and these have been criticized.

Researchers in the field of computer music initially tended to see MIDI as a "toy interface standard." Comparing the bandwidth of MIDI to, for example, DMA (Direct Memory Access) between a computer and dedicated peripherals is like comparing a garden hose to the Columbia River. But the years have shown that MIDI has a great deal of potential. Ultimately, the question about what we are giving up is less obvious than one might think.

Stored in the MIDI format, a fifteen-minute piece of music with sixteen voices will easily fit on an 800-K disk, tremendously reducing storage problems compared to digital audio — by around a thousand to one. Obviously, we have traded something here: We are limited to the sounds that the synthesizer or other devices we are controlling can make. Questions still remain as to whether MIDI is fast enough to capture every nuance, and whether two-way communication is possible? Though a great deal of thought was put into the standard initially, it was impossible to foresee just how far this standard would be pushed.

MAX is a visual programming language designed for music composition and performance. The musician who learns MAX can develop control software for MIDI-compatible synthesizers and related equipment. This software can be suited to a musician's individual needs or styles of working. Originally developed at the *Institut de Recherche et de Coordination Acoustique/Musique* (IRCAM) in Paris by Miller Puckette, MAX has been in use at computer music research facilities since 1988 and became a commercial product for the Apple Macintosh computer with the additional development of David Zicarelli in 1990. It is distributed by Opcode Systems, Inc., of Menlo Park, California.

If you connected a MIDI controller such as a keyboard directly to a sound-generating device, you would normally get a single sound for each note you play. However, by inserting a computer between the controller and the sound generator, it is possible to extend your playing to produce "more" than you are actually doing. The computer effectively reinterprets the meaning of the musical gestures produced by the musician. Simple examples of this reinterpretation might include sounding a chord when a single note is played, starting or stopping a melodic sequence, or transposing a melody to the key which corresponds to the note most recently played by a performer.

Because the possibilities of extending musical performance with a computer are so vast, the best approach to developing a piece that uses such techniques is an experimental and interactive one. Unfortunately, writing a computer program to send and receive MIDI commands is not a task which can be easily accomplished by most musicians. Even if one were to write such a program, it would be necessary to use a compiled language such as C or Pascal because the computer will be responding to the performer's gestures in real time and must execute rapidly. The need to simplify the process of developing MIDI control software motivated the development of MAX as a special-purpose language. MAX simplifies the task of MIDI software development in three ways.

First, since it is a visual language, people with little or no programming experience can learn the basic principles without months of study. MAX programs are called *patches* because the user "patches together" boxes (called *objects*) that represent input, output, arithmetic operators, or control functions. On the computer screen, the programs look like boxes connected by wires, so you can write programs just as you might configure a stereo system or use a modular synthesizer. Conceptually, one thinks of data as "flowing through" the wires from one box to another.

Second, MAX eliminates the need to deal with MIDI codes directly. One does not need to worry about the exact format of a MIDI message or how to send it to the serial port of the computer or, a task which is even more difficult, interpreting the type and nature of incoming MIDI messages arriving from a MIDI controller such as a keyboard. Instead, there are MAX objects for turning each part of a MIDI message into a *number* which is sent down a wire. For example, suppose our task is to write a MAX program that sends a MIDI note one octave above that which the computer receives. The MIDI note is received in an object called *notein*, which has three outputs, one for the pitch of the note, one for the velocity of the note (how hard it was played), and one for the MIDI channel (1 to 16). For the computer to merely repeat the note to the output, one just connects these outputs at the bottom of a *notein* object to the corresponding inputs at the top of a *noteout* object, as shown below. The noteout object takes care of putting the numbers back together into a MIDI message which is then sent out the serial port of the computer to a synthesizer.

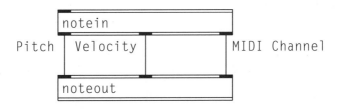

Now, to add an octave (12 steps) to the pitch value, we merely add 12 to the pitch value, leaving the other two numbers alone.

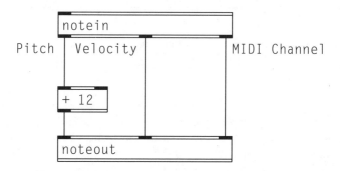

We could expand upon this example to modify the note's velocity or MIDI channel. In order to do this, we would deal with each of these numbers as an individual entity, because the MAX *notein* and *noteout* objects allow us to do so.

Finally, MAX contains a scheduler, a special facility for delaying an action until a specified time. This is especially useful in MIDI applications, because the MIDI specification says that you are responsible for turning off every note you turn on. Without a scheduler, after you turn a note on, your computer program would have to tie up the computer to wait until the specified time occurs to turn it off. With a scheduler, the programmer can specify when the note is to be turned off and the computer will regularly check the clock and execute the note-turning-off task at the specified time. Meanwhile, the computer can be used for other things, including turning on and off other events. In MAX, there are a number of different objects that make use of the scheduler. One of them is called *pipe*, so named because you can send a number into the pipe and it will come out after a certain delay. (In MAX, time is expressed in milliseconds, usually accurate enough for MIDI control tasks, though not appropriate for the actual synthesis of sound.)

Returning to our simple example above, we will modify the MAX program to delay an incoming note by one second by inserting a *pipe* between the input (*notein*) and the output (*noteout*). We need a separate pipe for each of the three numbers.

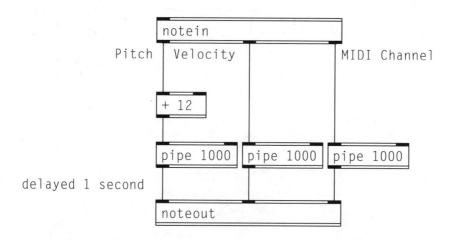

As an improvement, we can make two notes sound for each note played with a simple step. We merely reestablish the direct connection between the notein and noteout objects. This causes each number to go to two different places. The program shown at the top of page 255 will repeat the original note played on the keyboard, then, a second later, play another note an octave above the original. Note that we can make changes to the representation of the program on the computer screen without any intervening steps such as compilation. Thus, MAX lends itself to the experimentation that is so desirable when working with MIDI equipment.

Finally, here is an example of using a pipe to turn a note off after a specified amount of time. When the user clicks on the box labeled 60 (middle C), the number is sent immediately to three places (the order is

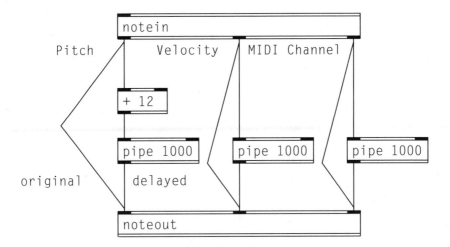

determined by the right-to-left position of the destination boxes on the screen). First, the number is sent to a pipe, which will delay it for 1000 milliseconds (one second). Second, it is sent to trigger another number, 64, which is in turn sent to the velocity inlet of the noteout object. Finally, 60 is sent directly to the pitch inlet of noteout, which sends the MIDI message out the serial port. A second later, the number 60 will come shooting out of the pipe, first sending a 0 to the velocity inlet of the noteout. When the piped 60 arrives at the noteout pitch inlet, the MIDI message to turn the note off is sent out. A MIDI note message with a velocity of 0 is interpreted to mean "turn the note at the specified pitch off."

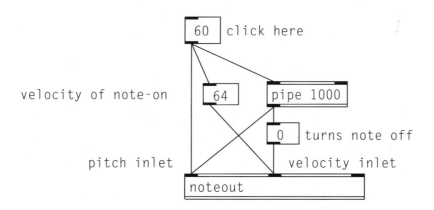

It turns out that most of the complexity of the example above has been incorporated into a single MAX object called makenote. At the top of page 256 is the same example rewritten to use this object, which is quite a bit simpler.

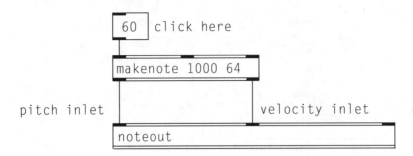

Conclusion

Because MAX makes it possible to work with individual numbers to control MIDI hardware rather than requiring the MIDI protocol, it has been possible to use a computer running MAX to combine MIDI instruments with other kinds of devices that do not "speak" MIDI. If a MAX object is created to support a new piece of hardware, it is usually quite trivial to incorporate it into a program. One can envision a situation in which hardware devices, as represented in MAX, are speaking a universal language of *numbers* which can be manipulated in arbitrary ways. The net effect can be a more musical style of control over equipment which might never have been considered in a musical context. For example, it is possible to use MAX to control the speed and direction of images on a video disk player from a MIDI controller. As MAX users continue to add to the language in the form of support for new devices, the possibilities for extending musical performance into new areas should continue to expand over the next several years.

Appendix I Bibliography

1. Compact Discs

Ultimately, sound is to hear, not to read about. Today the most enlightening and convincing publications concerning sound are digitally generated or processed sounds recorded on compact discs. Some appropriate discs are recordings of compositions using generated or processed sounds, such as the excellent Digital Music series on Wergo and others. There are also a few discs of psychoacoustic sound examples.

1-1. *Examples of Recorded Music That Illustrates Various Phenomena*

RISSET, JEAN-CLAUDE.   Excellent music. *Sud* (1985) is a wonderful example of the use of natural sounds, edited and combined with computer-generated sounds. In AC 1003, 1NA-GRM, Paris.

CHOWNING, JOHN.   Each of Chowning's compositions embodies something distinct and different. *Turenas* (1972) illustrates excellently the motion of sound through a room. Wergo, WER 2012–50. Schallplatten GmbH, Mainz, Germany.

1-2. *Psychoacoustic Examples*

MATHEWS, MAX V. (ED.)   *Sound Examples, Current Directions in Computer Music*, MIT Press, 1989. Compact disc. Distributed separately from the book *Current Directions in Computer Music* (Max V. Mathews and John R. Pierce, eds.), this compact disc contains a wide range of synthesized sounds illustrating various points. John R. Pierce wishes to apologize for the fact that one of the six examples he provided (75 through 81) is erroneous (number 76).

HOUTSMA, A. J. M., ROSSING, T. D., AND WAGENAARS, W. M. (1987).   *Auditory Demonstrations*. Philips 1126–061. Compact disc. This excellent eighty-track demonstration of various psychoacoustic phenomena comes with a booklet describing the demonstrations.

2. Periodicals

Examine journals in a library before subscribing. The following seem to the writer to be the most relevant journals.

AES, JOURNAL OF THE AUDIO ENGINEERING SOCIETY.   Audio Engineering Society, Inc., 60 East 42 Street, New York, N.Y. 10165-0075. For those interested in all the details of sound synthesis and reproduction.

MUSIC PERCEPTION. University of California Press, Department of Psychology, University of California, San Diego, La Jolla, Calif. 92093. Covers a wide range of musical phenomena.

COMPUTER MUSIC JOURNAL. MIT Press, 55 Hayward St., Cambridge, Mass. 02142. Quite wide coverage, with a good deal of the nitty-gritty of synthesis.

JOURNAL OF THE ACOUSTICAL SOCIETY OF AMERICA. Acoustical Society of America, 500 Sunnyside Boulevard, Woodbury, N.Y. 11797. This weighty journal covers all of acoustics. The papers on music and musical instruments, psychological acoustics, physiological acoustics, speech production, and speech perception will perhaps be of most interest.

3. Books

Books are cited in the hope that they can serve as a bridge to past publications and into ongoing activities. Some important papers published in various journals can be very useful. It isn't easy to select accessible and reliable books. Some get bogged down in details. Some cite out-of-date results or ideas. I advise looking at books in a library before purchasing them.

RANDEL, DON MICHAEL, ED. *The New Harvard Dictionary of Music*. Cambridge, Mass., The Belknar Press of Harvard University Press, 1986. A very useful volume.

BÉKÉSY, GEORG VON. *Experiments in Hearing.* Huntington, N.Y.: Robert E. Krieger Pub. Co. 1960, reprinted 1980. The work of a wise Nobel laureate, described by himself.

BREGMAN, ALBERT S. *Auditory Scene Analysis.* Cambridge, Mass: MIT Press, 1990. Our interpretation of sounds from the external world.

DEUTSCH, DIANA (ED.) *The Psychology of Music.* San Diego: Academic Press, 1982. Excellent. Includes chapters by bright people who are deep in computer music.

DODGE, CHARLES, AND THOMAS, JERSE. *Computer Music, Synthesis, Composition, and Performance.* New York: Schirmer Books, 1985. An excellent guide from excellent musicians.

FLETCHER, HARVEY. *Speech and Hearing in Communication.* New York: Van Nostrand, 1953. Still an interesting and useful book by a true pioneer.

GALILEO GALILEI. *Dialogues Concerning Two New Sciences.* Translated by Henry Crew and Alfonso de Salvio. New York: Dover, 1954. Pages 95–108 give Galileo's ideas concerning the relation of pitch (and harmony) to rate of vibration, including the way in which rate varies with the length, tension, and mass of a string.

HARRIS, CYRIL. *Handbook on Noise Control.* 2nd edition. New York: McGraw-Hill, 1979. By a truly reliable expert in architectural acoustics.

HELMHOLTZ, HERMANN L. F. *On the Sensations of Tone.* New York: Dover reprint, 1954. A must by the founder of modern psychoacoustics — and much more.

HUNT, FREDERICK VINTON. *Origins in Acoustics: The Science of Sound from Antiquity to the Age of Newton.* New Haven, Conn.: Yale University Press, 1978. The background that some unfortunately forget.

MATHEWS, M. V. *The Technology of Computer Music.* Cambridge, Mass.: MIT Press, 1969. Chiefly about software synthesis, but a fair amount about sound.

MATHEWS, MAX V., AND PIERCE, JOHN R. (EDS). *Current Directions in Computer Music Research.* Cambridge, Mass.: MIT Press, 1990. Twenty-one chapters by people active in computer music research.

MOORE, F. R. *Elements of Computer Music.* Englewood Cliffs, N.J.: Prentice-Hall, 1990. The author has been deep into digital sound generation and processing from almost the beginning.

PLOMP, R. *Aspects of Tone Sensation.* New York: Academic Press, 1976. A very informative book.

ROADS, CURTIS, AND STRAWN, JOHN (EDS.). "Foundations of Computer Music." Cambridge, Mass.: MIT Press, 1985. Revised and updated articles published from 1977 to 1979 in *Computer Music Journal* on digital sound synthesis, synthesis hardware, music software, and psychoacoustics and signal processing.

ROSSING, THOMAS D. *The Science of Sound.* 2nd edition. Reading, Mass.: Addison-Wesley, 1989. A compendium of all aspects of acoustics. The information on musical instruments is particularly helpful.

SABINE, WALLACE CLEMENT. *Collected Papers.* New York: Dover, 1964. How architectural acoustics come to be.

SCHUBERT, E. D. (ED.). *Psychological Acoustics.* Stroudsburg, Pa.: Dowden, Hutchinson, and Ross, 1979. The original papers reprinted here, which date from 1876 to 1970, give a wonderful sense of how the discoveries in acoustics were really made, and the wise introductory comments to the six sections give references up to 1977.

SUNDBERG, J. *The Science of the Singing Voice.* DeKalb, Ill.: Northwestern Illinois University Press, 1987. This careful student of singing tells a lot that is worth knowing.

Sources of Illustrations

page 3
Courtesy of the Conservatoire Royal de Musique, Brussels.

page 4
Courtesy of King Musical Instruments, Inc., Eastlake, Ohio.

pages 5 and 7 (top)
Music Division, The New York Public Library at Lincoln Center; Astor, Lenox, and Tilden Foundations.

page 6
© 1963 by Hermann Moeck Verlag, Celle. Used by permission of European American Music Distributors Corporation, sole U.S. agent for Hermann Moeck Verlag.

page 7 (bottom)
Courtesy of Lejarin Hiller.

page 9
Courtesy of Moog Music, Buffalo, New York.

page 10
Courtesy of James A. Moorer.

page 15
From *The Science of Sound*, by John Tyndall, Philosophical Library, New York, 1964.

pages 18–19
Adapted from a drawing in *Musical Acoustics: An Introduction*, by Donald E. Hall, Wadsworth, 1980.

page 20
Courtesy of the Metropolitan Museum of Art, Fletcher Fund, 1956.

page 23
From *Harmonie universelle*, by Marin Mersenne/courtesy of Martinus Nijhoff Publishers.

page 24
From *Anecdotal History of the Science of Sound*, by Dayton C. Miller. Copyright 1935 by Macmillan Publishers Co., Inc., renewed in 1963 by The Cleveland Trust Company, Executor.

page 25
Photography by Peter Simon/Black Star.

page 31
Photography by Saxon Donnelly/courtesy of the University of California, Berkeley.

page 33
Photography by Philip L. Molten.

page 43
Courtesy of Elizabeth A. Cohen, Andrew Schloss, and Eric Schoen.

page 45
Courtesy of the American Institute of Physics, Niels Bohr Library.

page 46
From *On the Sensations of Tone as a Physiological Basis for the Theory of Music*, by Hermann von Helmholtz.

page 48 (top)
Photography by Constantine Manos/ Magnum.

page 48 (bottom)
Photography by William P. Gottlieb.

page 49
Photography by Bob Shamis.

page 50 (bottom)
Adapted from a graph in "The Acoustics of Violin Plates," by Carleen Maley Hutchins. Copyright © 1981 by Scientific American, Inc. All rights reserved.

page 54
By A. M. S. Quinn/courtesy of Max V. Mathews.

pages 57 and 58
Courtesy of Elizabeth A. Cohen.

pages 65 and 66
Courtesy of Elizabeth A. Cohen.

page 77
Copyright © 1947 by Associated Music Publishers, Inc. All rights reserved. Used by permission.

pages 80 and 82
Adapted from graphs in *Experiments in Tone Perception*, by R. Plomp, Institute for Perception RVO-TNO, National Defense Research Organization TNO, Soesterberg, The Netherlands, 1966.

page 81
Adapted from a graph in "Critical Band Width in Loudness Summation," by E. Zwicker, E. G. Flottorp, and S. S. Stevens, *The Journal of the Acoustical Society of America* 29(1957):548.

pages 93 and 94
Courtesy of the Institut voor Perceptie Onderzoek, Eindhoven, The Netherlands.

page 99
From *Harmonie universelle*, by Marin Mersenne/courtesy of Martinus Nijhoff Publishers.

pages 103 and 104
Adapted from drawings in *Waves and the Ear*, by Willem van Berjeick, John R. Pierce, and Edward E. David, Jr., copyright © 1960 by Anchor Books. Reprinted by permission of Doubleday & Company, Inc.

page 105
Adapted from a drawing in "Neuroanatomy of the Auditory System," by R. R. Gacek in *Foundations of Modern Auditory Theory*, volume 2, J. V. Tobias, editor, Academic Press, 1972.

pages 108 and 109
Adapted from drawings in *Waves and the Ear*, by Willem van Berjeick, John R. Pierce, and Edward E. David, Jr., copyright © 1960 by Anchor Books. Reprinted by permission of Doubleday & Company, Inc.

page 112
From *The Wonders of Acoustics*, by Rodolphe Radau, Scribner's 1886.

page 113
Photography by Nobu Arakawa/Image Bank.

page 118 (top)
Adapted from a graph in *Modern Sound Reproduction*, by Harry F. Olson, Robert E. Krieger, 1978.

pages 121 and 123
Adapted from drawings in *Speech and Hearing in Communication*, by Harvey Fletcher, D. van Nostrand, © 1953 D. van Nostrand.

page 124
Reproduced from *The Acoustical Foundations of Music*, by John Backus, by permission of W. W. Norton & Company, Inc. Copyright © 1969 by W. W. Norton & Company, Inc., and John Murray, Ltd., London.

page 126
Adapted from a graph in "Musical Dynamics," by Blake R. Patterson. Copyright © 1974 by Scientific American, Inc. All rights reserved.

page 132
Adapted from graphs in *Speech and Hearing in Communication*, by Harvey Fletcher, D. van Nostrand, © 1953 D. van Nostrand.

page 135
Adapted from graphs in "On the Masking of a Simple Auditory Stimulus," by James P. Egan and Harold W. Hake, *The Journal of the Acoustical Society of America* 22(1950):622–630.

page 137
Adapted from a graph in "The Acoustics of Singing," by Johann Sundberg. Copyright © 1977 by Scientific American, Inc. All rights reserved.

page 148 (bottom)
Adapted from an illustration in *Music, Acoustics, and Architecture*, by Leo Beranek, copyright by John Wiley & Sons, Inc.

page 151
Adapted from a drawing in *The Collected Papers on Acoustics*, by Wallace Sabine, Harvard University Press, 1922.

page 152
Adapted from a graph in *Architectural Acoustics*, second edition, by K. D. Ginn, Bruel and Kjaer, 1967.

page 153
Photography by James R. Holland/Black Star.

page 155
Photography by Ezra Stoller, © ESTO.

pages 156–158
Adapted from graphs in "Acoustical Measurements in Philharmonic Hall," by M. R. Schroeder, B. S. Atal, G. M. Seeler, and J. West, *The Journal of the Acoustical Society of America* 40 (1966):434–440.

page 159
Photography by Susanne Faulkner Stevens.

page 161
Adapted from a graph in "Acoustical Measurements in Philharmonic Hall," by M. R. Schroeder, B. S. Atal, G. M. Sessler, and J. West. *The Journal of the Acoustical Society of America* 40 (1966):434–440.

page 163
Adapted from an illustration in "Construction of a Dummy Head after New Measurements of the Threshold of Hearing," by V. Mellert, *The Journal of the Acoustical Society of America* 51 (1960):1359–1361.

page 165
Courtesy of Manfred Schroeder.

page 166
Adapted from a drawing in "Binaural Dissimilarity and Optimum Ceilings for Concert Halls: More Lateral Sound Diffusion," by M. R. Schroeder, *The Journal of the Acoustical Society of America* 65(1979):958–963.

page 169
Adapted from illustrations in "Symposium on Wire Transmission of Symphonic Music and Its Reproduction in Auditory Perspective: System Adaptation," by E. H. Bedell and Iden Kemey, *Bell System Technical Journal* 13 (1934):301–308, © 1934 AT&T.

page 170 (top)
Courtesy of DAR Magazine.

page 170 (bottom)
Courtesy of Max V. Mathews.

page 176
Photography by Jacob/Black Star.

page 181
Courtesy of Ludwig Industries, Chicago, Illinois.

page 190
Adapted from a graph in "Electronic Simulation of Violin Resonances," by M. V. Mathews and J. Kohut, *The Journal of the Acoustical Society of America* 53 (1973):1620–1626.

page 192
Courtesy of Elizabeth A. Cohen.

page 193
Reproduced with permission from "Phonetics" in *Encyclopaedia Britannica*, fifteenth edition, © 1974 by Encyclopaedia Britannica, Inc.

page 198
Adapted from a drawing in "Scaling the Musical Timbre," by J. M. Gray, *The Journal of the Acoustical Society of America* 61(1977):1270–1277.

pages 200, 202, 203
Photographs courtesy of Ludwig Industries, Chicago, Illinois.

page 204
Photography by Michael Samson/courtesy of Amy Malina.

page 206
The Metropolitan Museum of Art, gift of Mrs. Alice Lewisohn Crowley, 1946.

page 216
©BEELDRECHT, Amsterdam/ V.A.G.A., New York, Collection Haags Gemeentemuseum — The Hague, 1981.

pages 242 and 243
Adapted from illustrations in *The Technology of Computer Music*, by Max V. Mathews, MIT Press, 1969.

Index

Page numbers in *italics* indicate illustrations.

265